Prentice Hall's Environmental Technology Series

Available:

Planned:

PRENTICE HALL'S
ENVIRONMENTAL TECHNOLOGY SERIES

Volume 3

Health Effects of Hazardous Materials

PRENTICE HALL'S
ENVIRONMENTAL TECHNOLOGY SERIES

Volume 3

Health Effects of Hazardous Materials

NEAL K. OSTLER

Salt Lake Community College

THOMAS E. BYRNE

Roane State Community College

MICHAEL J. MALACHOWSKI

City College of San Francisco

Prentice Hall
Upper Saddle River, New Jersey   *Columbus, Ohio*

Library of Congress Cataloging-in-Publication Data

Ostler, Neal K.
    Health effects of hazardous materials / Neal K. Ostler,
Thomas E. Byrne, Michael Malachowski.
        p.   cm. -- (Prentice Hall's environmental technology;
v. 3)
    Includes bibliographical references and index.
    ISBN 0-02-389551-9
    1. Toxicology.   2. Environmental toxicology.   I. Byrne, Thomas E.
II. Malachowski, M. J. (Michael J.)   III. Title.   IV. Series.
RA1211.086   1996
615.9--dc20

96-17137
CIP

Editor: Stephen Helba
Production Editor: Mary Harlan
Design Coordinator: Julia Zonneveld Van Hook
Text Designer: Custom Editorial Productions, Inc.
Cover Designer: Brian Deep
Production Manager: Pamela D. Bennett
Marketing Manager: Danny Hoyt
Illustrations: Academy Artworks, Inc.

This book was set in Utopia by Custom Editorial Productions, Inc., and was printed and bound by
Book Press, Inc., a Quebecor America Book Group, Inc. The cover was printed by Book Press, Inc.

© 1996 by Prentice-Hall, Inc.
Simon & Schuster/A Viacom Company
Upper Saddle River, New Jersey 07458

Printed in the United States of America

10 9 8 7 6 5 4 3 2 1

ISBN   0-02-389551-9

Prentice-Hall International (UK) Limited, *London*
Prentice-Hall of Australia Pty. Limited, *Sydney*
Prentice-Hall Canada Inc., *Toronto*
Prentice-Hall Hispanoamericana, S. A., *Mexico*
Prentice-Hall of India Private Limited, *New Delhi*
Prentice-Hall of Japan, Inc., *Tokyo*
Simon & Schuster Asia Pte. Ltd., *Singapore*
Editora Prentice-Hall do Brasil, Ltda., *Rio de Janeiro*

Contents

7
Toxic Agents 141

8
Environmental Toxicology 159

9
Radiation Health Effects 171

10
Common Chemical Hazards 183

Appendix A
Chemicals One May Encounter in the Workplace 209

Appendix B
Chemical Research Guide: Resources and Recommended Readings........ 219

Glossary 225

Index... 239

Preface

Goal of the Text

This text is the third in a planned six-volume series of Prentice Hall's Environmental Technology Series. It is written primarily as an introduction to the subject of toxicology for the hazardous materials technician and is a suitable textbook in an introductory course. Even though toxicology is a complex subject, this text has been written with simplified terminology and language accessible to the targeted audience and to the general public.

This text includes an introductory sample of a wide range of topics important to the hazardous materials technician. In brief, this book includes the following subjects:

- The origins of toxicology and a classification of toxic agents
- How toxic substances enter and are distributed through the body
- How the human body metabolizes toxic chemicals
- The effects of toxic substances on reproductive structures, resulting in mutations, cancers, and birth defects
- How toxic substances affect the major organs and organ systems of the body
- The groups to which various toxic agents belong and the target organs they affect
- Environmental pollutants found in our food, water, and air as well as a description of the dangers of environmental radiation
- Health effects of radiation
- Common industrial hazards

In addition to these subjects, Appendix A includes information on the health effects of common chemicals that might be encountered in the workplace. Appendix B lists some helpful reference books, government publications, periodicals, and electronic sources for further research on chemicals.

Development of this Text

In 1992, I developed the basic manuscript for this text through the tutelage of Harry Rollins, Medical Advisor to the Regional Poison Control Center of Salt Lake City and Director of the Center for Toxicological Studies at the University of Utah. In my early discussions with Dr. Rollins, I learned that there were several texts on toxicology written at the upper division and post-graduate levels, but none at the technician level. With Dr. Rollins' support and the contributions of several other authors with top credentials in the area of toxicology, we developed this textbook.

The coauthors made significant contributions to the writing of this text: Thomas Byrne of Roane State Community College in Tennessee and Michael Malachowski of The City College of San Francisco. Tom was involved as a contributing author to Volume 1 of this series. Michael and I have been associated as members of the Partnership for Environmental Technology and Education (PETE). He not only has taught the health effects course several times but also has developed a text of his own. Special thanks go to Doug Stutz of Miami-Dade Community College, whose work, widely known as the "Stutz Manual," is recognized as the standard reference for treatment of chemical exposure.

Acknowledgments

I would like to express my appreciation to George Van DeWater, originator of the Environmental Technology Program at Salt Lake Community College, who allowed me to complete the health effects course as a special study and who hired me as an instructor in the program. Thanks also to John T. Nielsen, co-editor of Volume 2, for the moral support he offered by participating in the series. Proper credit must also be given to the project editor, Stephen Helba, for believing in and supporting this project.

I wish to thank those who reviewed the manuscript: Kelly N. Bringhurst, Dixie College, and Dr. Norman R. Sunderland, Utah State University.

I must express my gratitude to my parents, A. Boyd and Lucille Peay Ostler. My father was a pharmacist, my grandfather a physician, and my uncle a noted ophthalmologist. Although all are now deceased, their influence on my life has led me to be at least on the periphery of medicine. I believe it would have pleased each of them to know of my involvement in the development and publication of this text.

Finally, I want again to acknowledge the understanding and support of my children who have sacrificed, along with me, the time this series of texts has taken. Robert was in junior high school when I began, and one Sunday afternoon this past August, we drove him to Snow College to enroll there and begin football practice. Dawn, Robert, Kimberly, and Patricia, I love and appreciate you very much.

Neal K. Ostler

About the Authors

Neal K. Ostler

Mr. Ostler is an adjunct faculty member of the Salt Lake Community College where he lectures on environmental technology and coordinates the Environmental Training Program. He is a graduate of the University of Utah and is a Certified Hazardous Materials Manager (CHMM). Mr. Ostler is certified as a "Train-the-Trainer" in a variety of subject areas and provides SARA/OSHA training as the contract instructor for the Industrial Training Institute.

Mr. Ostler has numerous years experience in law enforcement. While working for an alcohol and substance abuse program at the Utah State Prison, he developed a biblio-therapy program for inmates housed in maximum security. He also is experienced in hazardous materials emergency response. Neal is a graduate of EPA's Environmental Crimes Investigation Training and has attended over 1,000+ hours of environmental/hazardous materials workshops and seminars.

Thomas E. Byrne

Thomas E. Byrne resides in Harriman, Tennessee, and teaches at Roane State. He received his B.S. and M.S. from Tennessee Technological University in biology, and Ph.D. from the University of Tennessee at Knoxville in botany and microbiology. He is a registered microbiologist (RM-AAM) with the American Academy of Microbiology, and is a member of the American Society of Microbiology, Environmental Microbiology Section.

Dr. Byrne has served as a member of the Executive Committee of the Tennessee Academy of Science and is a life member of the Academy. He has authored many articles for refereed journals, as well as two laboratory manuals. He also serves as a consultant to the Oak Ridge National Laboratory (ORNL) in the area of viral carcinogenesis and radiation assessment and has administered radiation technician training programs for Martin Marietta Energy Systems in Oak Ridge, Tennessee, and Paducah, Kentucky. His most recent consulting duties include environmental radiation assessment at ORNL for the Nuclear Regulatory Commission.

M. J. Malchowski

Dr. Malachowski received his Ph.D. in biophysics and health science from the University of California, Berkeley, in 1978; he specialized in space radiation biology. He was a Lawrence Berkeley Laboratory Guest from 1972 to 1979, a Research Associate at NASA Ames Research Center from 1975 to 1979, and a Stanford Synchrotron Radiation Laboratory Guest from 1977 to 1980. He performed postgraduate work at the University of Western Ontario in microchemical spatial analysis on biological organisms and was a Fellow of the Canadian Medical Research Council from 1977 to 1980.

Dr. Malachowski started the Solid and Hazardous Waste Management program for the Peralta Community Colleges in 1980, creating courses in Toxicology and Environmental Toxicology. Since 1981, he has taught Anatomy and Physiology at City College of San Francisco; he developed the curriculum and teaches in their Environmental Technologies program. He has been working for San Francisco State University since 1992 to promote environmental awareness in the San Francisco and East Bay communities. He was a Partnership in Environmental Technology Education (PETE) associate to the Navy in 1994 and to the California Department of Toxic Substance Control (DTSC) in 1995. Dr. Malachowski performs consulting in a number of areas. His firm, CCE, has received five Small Business Innovative Research (SBIR) awards from DOE, DOD, and NASA.

Douglas R. Stutz

Douglas R. Stutz, Ph.D., currently serves as a Professor of Environmental Science Technology for Miami–Dade Community College, Miami, Florida. Dr. Stutz has wide experience in the fields of hazardous materials management and chemical casualty management. During a career with the Surgeon General's Office of the United States Army, he served as Executive Officer, Walter Reed Army Institute of Research; Director, Microbiology Laboratory, SEATO Medical Research Laboratory, Bangkok, Thailand; Professor of Emergency and Contingency Medicine, Uniformed Services University of the Health Sciences; and other positions dealing with the medical effects of various exposures. He then became a Fire Commissioner in Maryland, and subsequently, Fire Chief, Miramar, Florida. Dr. Stutz has written or coauthored over 20 books and more than 100 articles for refereed journals, magazines and newspapers. Currently, he is the editor of the Florida *EMS Journals,* writes for *The Firefighter News* on the subject of hazardous materials, and is the editor/publisher of an environmental newsletter. Dr. Stutz has taught extensively within and outside the United States. His consulting firm, GDS Communications, Inc., provides seminars on the subject of hazardous materials.

Introduction

SCOPE AND BRIEF HISTORY OF TOXICOLOGY

Every year in the United States, about two million people are poisoned. This number includes industrial poisonings, poisonings at work and at home, accidents, and suicides. Because of the high incidence of poisoning, intentional or unintentional, the government has created a national network of poison control centers.

Information concerning poisons and poisonings can be obtained from the Poison Control Centers throughout the United States, which offer a variety of services. For example, they can supply information on the ingredients of many compounds bearing trade names, the potential dangers from such compounds, and measures for handling emergencies arising from the ingestion of these or other toxic substances.

Many of the centers maintain a telephone information service 24 hours a day. They are usually associated with local departments of health or a community hospital and can be located through these agencies. For example:

Rocky Mountain Regional at Salt Lake City, UT (800) 456-7707
UC San Francisco Poison Control Center (800) 233-3360

In industrial settings, employers are required to maintain a file of information on the chemicals found or used at a job site. Employers are required to provide employee access to Material Safety Data Sheets (MSDS) containing the required information at any job site. The MSDS is a summary of the important physical characteristics, health, safety, and toxicological information on the chemical or substance. MSDS sheets are supplied by the chemical manufacturer, for all their hazardous materials, as mandated by the Hazardous Communication Standard.

The U.S. Department of Labor, Occupational Safety and Health Administration (OSHA) determines what must be listed on the material safety data sheet. (If you have to refer to specific federal requirements, the laws are collected in the Code of Federal Regulations [CFR]; OSHA's Hazardous Communications Standard is found in 29 CFR 1910.1200.)

Human interaction with hazardous and toxic chemicals has occurred throughout recorded history. You may be familiar with the 19 recent industrial poisonings involving methyl isocyanate in Bhopal, India, that resulted in many deaths.

Historical references to industrial poisonings are many. For example, Hippocrates (370 B.C.) described abdominal colic in a man who mined and extracted metals. Theophrastus of Erebus (370–287 B.C.) and Pliny the Elder (A.D. 23–79) wrote of human interactions with arsenic and mercury.

Throughout history, artists and craftsmen have worked with toxic chemicals. Arsenic was used for decorations in Egyptian tombs; cadmium was used to make a red paint for art works. Lead was commonly used in paints and glazes for pottery. Lead contamination of wine has been postulated as leading to the fall of the Roman empire.

Even our language is sprinkled with terms derived from the study of poisons. The phrase "mad as a hatter" derives from the fact that hat makers were subject to the effects of mercury toxicity because they used mercury—a heavy metal that impairs brain function—in the processing of felt.

Even before there were "industrial uses," many toxic chemicals were already known to the ancients who were aware of the toxic effects of poisonous plants and snake venoms. Ancient people used this knowledge in hunting (curare on arrow tips) and in the practice of medicine (plant alkaloids) to treat various conditions. The Chinese, Aztecs, and Mayans had extensive collections of herbal remedies.

It has long been known that some substances are deadly poisons and they were used to kill. Socrates, the Athenian philosopher and teacher, took his life by drinking poisonous hemlock. Catherine de Medici, a member of the Borgia family of Italy, used poisons for personal gain and financial profit in the Middle Ages.

Shakespeare appeared to have extensive knowledge of poisons. The Shakespearean plays of *Hamlet* and *Romeo and Juliet* refer to poisonings common for the time. In *Hamlet,* poison is placed in the king's ear to kill him. Hamlet himself was killed with a sword tip that had been dipped in poison. In *Romeo and Juliet,* poison is ingested. Human interaction with toxins probably occurred long before recorded history and is associated with almost all cultures.

Paracelsus, Phillippus Aureolus Theophrastus Bombastus von Hohenheim, (1493–1541) was a Swiss physician who is credited as the father of toxicology. He was a free thinker who wrote in German rather than Latin. Although he was disdained by his medical peers, he was very popular with the gypsies and tradespeople.

Paracelsus, far ahead of his time, provided a bridge between alchemy and science. He frequently attacked and ridiculed the traditional and venerated ideas and institutions because he believed that they were erroneous or baseless superstition. He held most of the medical doctrines and methods of the day in contempt. His views earned him the disdain of the esteemed medical establishment, which led to numerous moves of his medical practice. He died at the age of 48 from wounds he suffered in a tavern brawl.

▶ **TABLE I–1**

Approximate oral LD_{50} values in rats. As first expounded by Paracelsus, while the correct amount of a substance may be required for health, too much of almost anything may be toxic. The amount of a substance required to kill you is termed the Lethal Dose (LD). An LD_{50} (mg/kg) is the amount of a substance, dose, per kilogram (2.2 pounds) of body weight necessary to kill 50% of a particular population. This table provides some values for (1) substances necessary for health, such as sodium chloride and vitamin A; (2) substances that are useful or desirable, such as sucrose, alcohol, aspirin, and caffeine; and (3) substances that are harmful or toxic, such as nicotine, cyanide, and strychnine. The table values are those for rats; a similar range of toxicity might be expected for humans.

Agent or Chemical	LD_{50} (mg/kg)
Sucrose (table sugar)	29,700.0
Ethyl alcohol (grain alcohol)	14,000.0
Sodium chloride (table salt)	3,000.0
Vitamin A	2,000.0
Vanillin	1,580.0
Aspirin	1,000.0
Chloroform	800.0
Caffeine	192.0
Phenobarbital (barbiturate)	162.0
DDT	113.0
Sodium nitrite	85.0
Nicotine	53.0
Sodium cyanide	6.4
Strychnine	2.5

Paracelsus was most famous for his contention: "All substances are poisons; there is none which is not a poison. No substance is a poison in itself, it is the dose that makes a substance a poison. The dose differentiates a poison and a remedy. The dose makes the poison." (See Table I–1.) He distinguished between acute and chronic toxic effects of metals and described, in detail, the symptoms of chronic exposure to mercury. In 1567, 26 years after his death, Paracelsus's monograph on occupational diseases of miners and smelters was published.

A thorough and systematic study of toxins, by chemical and biochemical analysis and laboratory testing, is a relatively recent undertaking. Federal and state governmental protection from toxins in the workplace is relatively recent, and standards are continually being updated as more scientific data and information become available. New limits are published yearly. The National Institute of Occupational Safety and Health (NIOSH) annually publishes a "Registry of Toxic Effects of Chemical Substances."

A toxic substance can be defined as follows:

A toxic substance is any substance or material that can:

▶ Induce cancer, tumors, or serious tissue disorders in people or in experimental animals
▶ Induce a permanent change in the characteristics of an offspring from those of its parents, either human or animal

▶ Produce physical defects in the developing human or experimental animal embryo
▶ Produce death in animals exposed via the respiratory tract, skin, eye, mouth, or other routes in experimental or domestic animals
▶ Produce irritation or sensitization of the skin, eyes, or respiratory passages
▶ Diminish mental alertness, reduce motivation, or alter behavior in humans
▶ Adversely affect the health of a normal or disabled person of any age or of either sex by producing reversible or irreversible bodily injury or by endangering life or causing death from exposure via the respiratory tract, skin, eye, mouth, or any other route in any quantity, concentration, or dose reported for any length of time.

As a scientific area, *toxicology* is the study of poisons, their actions, their detection, and the treatment of the conditions produced by them. An individual skilled in the science of toxicology is referred to as a *toxicologist.*

The quality of being poisonous is called *toxicity,* and the science expresses this by a fraction indicating the ratio between the smallest amount that will cause an animal's death (lethal dose) and the weight of that animal (generally, in kilograms, kg). The condition of general intoxication due to the absorption of toxic products is called *toxemia.*

Additional terminology used throughout this book includes:

▶ *Toxinemia,* a poisoning of blood with toxins
▶ *Toxicosis,* any disease condition due to poisoning
▶ *Toxipathic,* pertaining to the pathogenic action of toxins

The purpose of this text is to present an objective discussion of what makes chemicals toxic to humans. This is for people who want a real understanding of the concepts of chemical exposure. It is expected that the reader or student of this text will become familiar with the laws of nature that govern chemicals and will better understand the threats or risks of living in a world of chemicals.

1

Principles of Toxicology

COMPETENCY STATEMENTS

This chapter introduces the principles of toxicology. After reading and studying this chapter, students should be able to meet the following objectives:

▶ Define the meaning of and describe three common approaches to the subject of toxicology.

▶ Demonstrate a basic understanding of the following components of toxicity: (a) sites and routes of entry, (b) chemical concentration, (c) dose, (d) duration, and (e) frequency.

▶ Identify and discuss the different factors that affect the total concentration of chemicals in the body.

▶ Identify and discuss the personal differences that affect the toxicity of chemicals on the individual.

▶ Describe and discuss the differences among (a) additive effects, (b) antagonistic effects, (c) synergistic effects, and (d) potentiation.

▶ Identify the meaning of *tolerance* as it relates to chemicals that are foreign and harmful to the body.

▶ Describe the important relationship of dose and response.

▶ Identify important protocols for the toxicity testing of laboratory animals.

▶ Discuss the concepts of risk assessment, safety factors, and risk extrapolation.

INTRODUCTION

In this chapter, students will be introduced to the basic study and concepts of toxicology, or the science of poisons. Since antiquity, human beings, like all living organisms, have been concerned with survival. A survival mechanism adopted by numerous plant and animal species is to contain or distribute poisons or noxious substances. Thus, they can avoid being eaten or becoming prey to others. Such species are typically avoided as food sources, and initially, humans learned to avoid these species. However, being clever and ingenious, they went on to use these poisons as, for example, aids to hunting. By tipping an arrow with a poisonous substance, it was necessary only to hit the target with the arrow; the poison, not the arrow, killed or incapacitated the prey.

The Greek word *toxikos* means "of or for a bow" and originally had nothing to do with poison. But the use of poison on arrow tips soon made people associate the word *toxon* with "poison," leading to our current use of the word *toxic,* or "poisoned." The Greek suffix *-ology* means "the study of." Therefore, *toxicology* is the study of poisons, or the effects of chemicals, on the body.

TOXICOLOGY

Toxicology is a subdivision of pharmacology, the study of the effects of chemicals on the body and the response of the body to chemical substances. Toxicology is an extremely broad science because of the variety of adverse effects attributable to toxins and the diversity and pervasiveness of the chemicals present in our environment. Consequently, the field of toxicology is frequently divided into a number of subfields. Common divisions of toxicology that are briefly discussed in this chapter are descriptive toxicology, mechanistic toxicology, and regulatory toxicology. Toxicologists frequently specialize and practice in one of these subfields.

Descriptive Toxicology

Descriptive toxicology is concerned with the quantitative and qualitative analysis of the adverse effects of different substances. *Quantitative analysis* concerns the study of the amount of a poison necessary to cause harm, whereas *qualitative analysis* deals with the types of harm caused by a poison.

Descriptive toxicology seeks results and information that can be used to assess risk to humans and other living organisms. The risk to humans might be through the use of drugs, one's normal diet, or by accidental exposure to chemicals in the workplace or at home. The risk to others includes the risk to birds in the air, fish in the waters, plants on the land and in the water, and other organisms or microorganisms, which might disturb the balance of the ecosystem.

Mechanistic Toxicology

Mechanistic toxicology is concerned with determining the mechanisms by which poisonous chemicals exert a toxic effect on living organisms. It looks at how chemicals get into the body, how they are distributed, what effects they have on the various organs, how the body attempts to defend itself, and, finally, what mechanisms the body uses to remove these chemicals. Typically, this area is divided into the areas of pharmacokinetics and pharmacodynamics. Because these areas deal specifically with toxic effects, they are also known as toxicokinetics and toxicodynamics.

Toxicokinetics or *toxikinetics* deals with the absorption and movement of toxic substances by and through the human body. After a substance is absorbed, it is distributed throughout the body by the bloodstream or circulatory system. As the chemicals are distributed, they may have adverse or harmful effects on the different systems of the body such as the central nervous system, the reproductive system, or the immune system. The physical state (solid, liquid, or gas) and chemical properties of a substance affect how it gets into the body (route of entry) to be absorbed into the bloodstream, how it is accumulated in the different tissues and organs, how it is transformed into other chemicals and substances, and how it is removed or excreted. One's physical and mental health can also greatly affect all of these kinetic processes.

Toxicodynamics is the study of the actual events that occur in the body when it is exposed to a certain toxic substance; it is the measurement of the cause and effect on the body as the substance is distributed to the various tissues and organs. In many cases, the toxic substance affects specific sites or groups of sites or locations in the body. The substance may produce a variety of events at that specific site because it interacts with other materials present in the blood or in the tissues. The body reacts to the substance differently at different sites because of biochemical processes. Such processes lead to the beginning of a stimulus in that site that can be measured by body symptoms, and the site can be identified as a "target organ" of that particular chemical. In most cases, the arrival of the substance evokes events that result in the final effect, which is measured by the toxicologist. Generally the magnitude of biological action, or reaction to the chemical's presence, is directly related to the concentration of the toxic substance at the target site.

The study of the effects on different systems of the body is discussed at length in Chapter 5, "Systemic Toxicology."

Regulatory Toxicology

Regulatory toxicology concerns the decisions of establishing "safe" levels of exposure, first by investigating how a chemical poses a danger or risk and then by determining which dosages might constitute a risk for particular sets of organisms subject to exposure. Subsequently, a "safe" level is determined for regulatory purposes. Many government agencies have departments of toxicology involved in establishing standards for the amounts of chemicals to be permitted in our environment.

RISK VERSUS SAFETY

To be practical, the critical factor is not the toxicity of a substance but the risk or hazard associated with its use. A human hazard can be defined as any event or circumstance that threatens or jeopardizes the health or safety of an individual or a population group such as exposure to chemicals. Risk, meanwhile, is the probability of the event or circumstance occurring and depends on the following factors: the

▶ **FIGURE 1–1**
The hazard associated with
a particular substance depends on the three elements
represented by the sides of
the triangle.

organism, the poison, and the environment. These factors are called the "toxic tri-angle" (Figure 1–1).

While risk is the probability that a substance will produce harm under specified conditions, safety is the probability that no harm will occur under those conditions. Therefore, depending on its use and handling, an extremely potent, supertoxic chemical may not be as hazardous as a less potent, toxic substance: it all depends on the actual exposure, absorption, and dosage of the chemical.

What constitutes an acceptable risk is a matter of judgment. These decisions are complex and multifaceted and involve an analysis of benefit versus risk, and cost versus benefit. High risks, which may be acceptable in the use of life-saving drugs, would obviously not be acceptable for use as food additives.

Factors evaluated when determining acceptable risk include:

1. Benefits gained from use of the substance
2. Adequacy and availability of alternatives
3. Anticipated extent of public use
4. Employment (worker) considerations
5. Economic considerations
6. Effects on environmental quality
7. Conservation of natural resources
8. Public opinions
9. Social values, such as environmental justice

CLASSIFICATION OF TOXIC AGENTS

Because of the large number of toxins and potential poisons that exist, it is advanta-geous to divide them into groups or sets. This process of segregation and division is known as *classification.* Toxic agents may be classified in any of a variety of manners. The choice of groupings depends on the particular needs of the situation.

Because this text is concerned with how toxins affect the body's organs or organ systems, we will classify them in terms of their target organ, such as hepatotoxins, which target the liver. We are also concerned with the particular sets of toxic health effects, and so we will also classify chemicals by these effects, such as carcinogens, which are those that cause cancer. Additionally, we can classify chemicals according to their physical state (solids, liquids, and gases), their poisoning potential, their chemistry, their labeling requirements, their general physical behavior, and their fre-quency of behavior in the workplace. In this text, the reader will discover a variety of groupings and classifications of toxic substances.

THE COMPONENTS OF TOXICITY

Exogenous chemicals are those that are foreign and harmful to the body. While toxicity is the measurement of the body's adverse response from exposure to these exogenous chemicals, a number of parameters must be determined to fully characterize the toxicity of a specific chemical.

In a sense, the toxicologist needs to determine the answers to pertinent questions much like a police officer who is investigating an incident: The officer needs to discover where the incident occurred, how the perpetrator made entry, and how many perpetrators there were. Thus, the toxicologist must have answers to the following questions:

▶ Where and how did the chemical get into the body?
▶ What was the quality of the chemical to which the person was exposed?
▶ How large a dose was required to produce the effect?
▶ How long was the duration of exposure?
▶ How often did the exposure occur?
▶ What is the total chemical burden of the body?
▶ Who or what is the subject of the study?

In the following section we will briefly discuss each of these items, and several of them will be discussed in more depth later in the chapter or elsewhere in the text.

Sites and Routes of Entry

Where and how did the chemical get into the body? Our body is equipped with a protective membrane, the skin (integumentary system), which is the body's natural barrier to the outside world. Exposure begins with the process in which a substance is presented to this protective membrane. Certain areas of this membrane are more vulnerable than others because of openings in the membrane that accommodate oil and sweat glands, hair, and orifices for other body functions. Further information is also needed about particular routes of entry, and although they will be discussed in Chapter 2, they are briefly discussed here.

Inhalation

Inhalation is the most common route of exposure in an industrial setting or at the workplace. Substances that contaminate the air we breathe are drawn into the lungs, absorbed by the blood, and quickly distributed throughout the body via the circulatory system. Many corrosive substances also produce local damaging effects to the surface of the respiratory tract such as oleum, chlorine, hydrochloric acid, and sulfur trioxide.

Dermal Absorption

Skin contact, or *dermal absorption,* is fairly frequent with the hands and, depending on the job and substance being used, many other surfaces of the body. Washing items in solvent-based solutions with bare hands is a common method of exposure. Spraying pesticides will frequently expose the body's surfaces to a mist, which may be absorbed through the skin. Contaminated clothing that remains in contact with the skin offers continued exposure and absorption.

Ingestion

Ingestion, or swallowing, is a very effective route into the body, and most drugs are given orally for just this reason. This method of entry is not a great problem in the industrial environment because, if workers keep their mouths shut, it is difficult for substances to gain entrance to the digestive tract. Ingestion also occurs from the swallowing of mucus that has been delivered to the mouth via the mucociliary transport. All too frequently, however, poor hygiene, smoking, chewing of gum, and eating with unwashed hands lead to exposure by ingestion.

Injection

For the fullest and quickest response, the most efficient route of entry into the body is by direct injection into the bloodstream. This can occur by stepping on rusted nails or cutting oneself with some other sharp object that is contaminated. Airless paint sprayers are another means of injecting substances directly into the bloodstream.

Eye Hazards

A very efficient mechanism as a route of entry into the body is via the moist surface of the eyes, which offer a ready environment for dissolving substances. The optic nerve is a direct path to the central nervous system, and the eye is also rich in blood vessels offering ingress to the circulation to other systems of the body as well.

Routes of exposure are discussed in more detail in Chapter 2.

Chemical Concentration

What was the quality of the chemical to which the person was exposed? This question is not one regarding the identity of the chemical but rather the quality or concentration and includes such issues as potency and efficacy. The issue of concentration also involves questions about the physical state of the chemical substance, including its form or state of matter (Is it in granules or powder solids? Is it a liquid? Was it a very fine vapor mist or minute solid fume?). The smaller the form, the easier it will be to make entry into the body.

Concentration relates to how pure a substance is: if it is not diluted, then it is said to be of 100% concentration. The values for the toxicity rating assume 100% purity or concentration.

Frequently, in comparing the toxicity of two or more chemicals, the terms *potency* or *efficacy* may be used. Potency refers to the range of doses that produce increasing and more lethal responses. Efficacy, on the other hand, expresses the limits of the relationship between dose and response.

Potency may be expressed on the "Toxicity Rating Scale" shown in Table 1–1.

Efficacy refers to how well a substance does its job. For example, taking an aspirin for an upset stomach is not effective; in fact, such a treatment is likely to upset the stomach even more. The potencies of the aspirin and ibuprofen for relieving a headache, however, may be compared to each other; that is, is one pill of aspirin more effective than one of ibuprofen? However, to make this comparison, the dose required to relieve the particular symptoms must also be considered. If the aspirin pill is 500 mg (milligrams) while the ibuprofen is 200 mg, and they both work equally well, then the ibuprofen is identified as the more potent. The two treatments, however, have equal efficacy because they each cure the headache.

▶ TABLE 1–1
Toxicity rating scale for 175-pound (79 kg) person.

EPA #	Oral LD$_{50}$	Toxicity	Metric and English Units
6	< 1.0 mg/kg	Dangerously	70 mg = mist
5	1–50 mg/kg	Highly toxic	350 mg = 3 drops
4	50–500 mg/kg	Toxic	3.5 grams = .1 oz
3	500–5,000 mg/kg	Moderately toxic	35 grams = 1 oz
2	5–50 gram/kg	Low toxicity	350 grams = 10 oz
1	> 50 gram/kg	Non toxic	3,500 grams = 1 gal

Dose

How large a dose was required to produce the effect? This question regards the quantity of chemical that is absorbed and available for metabolic interaction and is discussed in-depth on p. 16.

Duration

How long was the duration of exposure? Establishing duration is not always easy, but it is a very important task to establish the full nature of the exposure. When working in an environment where people may be forced to breath contaminated air, the duration of exposure is fairly easy to time. Splashing a substance in the eyes and then flushing it with water, however, is an exposure period that is more difficult to quantify, as is swallowing a substance or injecting it under the skin in an accident, such as tripping and falling.

The basic subcategories of duration range from acute to chronic. Between the two extremes is an infinite variety and number of combinations that can occur with different mixes of frequency (how often), quantity or dose (how much), and duration (how long).

The different categories and subcategories included in our discussion are acute, subacute, subchronic, chronic, and long-term exposures.

Acute Exposure

This exposure occurs when rapid symptoms or health effects may be demonstrated from a relatively small number of doses and during a short period of time. An acute exposure is often measured in terms of one exposure (single event) during a very brief period of time (duration), which is usually less than 24 hours. An example of an acute exposure is when a substance is spilled and people working in the immediate area are exposed. A chemical is said to have acute toxicity when it has an ability to cause systemic damage, noticed symptoms, as a result of a one-time exposure, a single event.

Subacute Exposure

This refers to 13 to 40 exposures over a period of several weeks to a month.

Subchronic Exposure

This refers to 30 to 90 exposures over a period of one to three months.

Chronic Exposure

This refers to more than 90 exposures for periods in excess of three months. Chronic toxicity refers to the ability of many repeated exposures to low levels of the chemical over a relatively prolonged period of time to cause systemic damage. Examples of chronic effects include the impact on health by pesticide residues and minor but repeated exposures in the workplace. Another example of a chronic exposure is that of a mechanic washing auto parts in solvent, which may result only in a slight irritation of the skin at first, but with multiple occurrences each day over a lifetime, may cause central nervous system problems and serious skin disorders.

Long-Term Exposures

These are exposures to very small doses but occur at a high frequency over a period of years. This generally refers to a living situation, such as a young child living in a house contaminated with lead dust, as opposed to industrial exposures.

In most cases, a correlation between acute and chronic effects does not exist: the two extremes of exposure affect different organ systems, and chronic effects cannot be predicted from a knowledge of the effects produced by an acute exposure. An example would be that of acute arsenic intoxication, which results in profuse and painful diarrhea, vomiting, and other gastrointestinal disorders, whereas chronic exposure leads to skin changes and damage to the liver and central nervous system.

Frequency

How often did the exposure occur? Frequency is the number of exposures in the time frame and is often more significant than duration or quantity over the long term. Repeated exposures to many mildly toxic substances can lead to compromise and eventual breakdown of the targeted organ system.

Concentration in the Body, or Total Body Burden

What is the total chemical burden of the body? Both the duration and the frequency of administration of the dose are critical time-related factors that must be considered when trying to determine the "body burdens," or the actual amount of toxin in the organism. Another way of seeking this information would be to ask, "What is the total concentration of the chemical in the organs of the body?"

The following factors affect concentration in the body organs:

1. Barriers may prevent absorption of the toxin into the bloodstream. The size of the molecule as compared to size of the cell membranes is important. The nature of the molecule is also important. Is it water soluble or fat soluble?
2. Distribution is determined by the rate of blood flow to the target tissues. It also reflects the ability of molecules to cross over into the tissues from the capillary bed.
3. Storage sites are locations such as blood plasma, liver, kidney, fat tissue, and bone where molecules may be captured and stored.
4. Excretion is the removal of molecules by urine (kidney), bile (liver), exhalation (lungs), feces (GI tract), cerebral fluid (brain/blood), sweat and saliva, and regurgitation from the stomach.
5. Absorption is the reverse process of excretion and occurs during exposure.

6. Biotransformation is the body's process of chemically converting toxic chemicals to allow excretion and destruction. It varies according to an individual's age, gender, genetics, health and nutritional status, and disease state.

Personal Differences

Who or what is the subject of the study? Individual differences are characteristics that vary from person to person and include race, gender, age, health, disease state, physical condition, genetics, and basic personal immunity. Each will be discussed in more detail in Chapter 3.

For further discussion of the time-base relationship, see "A Closer Look" at the end of this chapter.

Remember, the study of the toxicity of a chemical requires a diverse set of information, and putting all of the pertinent pieces of information together is a complex study, an investigation.

UNDESIRED EFFECTS

Although death is a common "end point" used to evaluate the potency of particular toxins, a number of other negative effects are also caused by toxins: allergic reactions, idiosyncratic reactions, immediate versus delayed toxicity, reversible versus irreversible effects, and local versus systemic toxicity.

Allergic Reactions

The body's immune system is activated by a variety of chemical agents in a number of ways. Frequently, these adverse reactions to a chemical occur even at very low doses. The symptoms and manifestations are numerous and may involve systemic responses. They range in severity from minor skin irritations to anaphylactic shock and death.

Idiosyncratic Reactions

These reactions occur because of specific genetic predispositions within small segments of the population. These reactions may range from an extreme reaction to low, generally nontoxic doses to extreme resistance and insensitivity to high, generally toxic doses of the same drug or chemical.

Immediate Versus Delayed Toxicity

Immediate reactions are those that develop rapidly after administration of a single dose. Delayed reactions are those that do not develop until after a lapse of time. For example, carbon monoxide inhalation causes immediate loss of consciousness, whereas drinking alcohol requires some period of time before intoxication occurs.

Reversible Versus Irreversible Effects

Sublethal doses, that is, doses smaller than that required to kill, are frequently reversible; once recovered, subjects are no worse off than they were before the exposure. An example would be a person's complete recovery from a hangover. In other situations, such as in carbon monoxide poisoning, if the subject was unconscious and deprived of oxygen for too long, organic damage—damage to the brain—could occur, making full recovery impossible.

Local Versus Systemic Toxicity

Effects may be characterized as local or systemic. Poison oak or ivy exhibit local effects, such as a rash at the site of exposure. In general, substances that affect topically—at the site of application—are considered local. Most corrosive substances, such as acids or bases, exhibit local effects because their corrosive actions typically damage the cells that they contact. Therefore, the actions of corrosives are also irreversible. Other substances produce systemic effects. Systemic substances are first absorbed into the body and then circulated around the body via the circulatory system. Because of this dissemination, they can affect entire body systems, hence the term *systemic*. However, most substances do not affect the entire body; rather, they are organ or system specific, affecting only the target organ.

CHEMICAL INTERACTION

When a person simultaneously takes more than one drug, the combination may or may not alter the effects of each chemical.

Additive Effects

If two or more chemicals or substances act together to produce an effect that is equal to the sum of the individual effects, it is said to be *additive*. An example is that the sum of the effects of taking one Anacin and one Tylenol (1 + 1 = 2) is similar to the effects of taking two Anacins or two Tylenols (2 = 2). Doses of chemicals such as epinephrine (adrenalin) and norepinephrine (noradrenalin) tend to be additive.

Antagonism

An *antagonistic effect* occurs when two or more chemicals interfere with one another and the sum of their combined effects is less than the total of individual effects (2 + 2 < 4). There are four major types of antagonism, and it is important to distinguish them as they form the basis by which antidotes are determined.

1. *Functional:* This type of antagonism produces a smaller or lesser effect by competition for the same effector site. Examples of direct competition for binding are the treatment of cyanide poisoning with injections of hydroxocobalamin and dicobalt edetate, treatment of fluoride burns with calcium gluconate, and chelation therapy with EDTA for lead poisoning (also discussed in Chapter 3).
2. *Chemical:* This type of antagonism occurs when the different chemicals have opposite effects that actually inhibit each other. An illustration of an antagonism acting directly to influence enzyme activity is the treatment of cyanide poisoning by injection or oral ingestion of sodium nitrate, sodium thiosulfate, or methylene blue.
3. *Dispositional:* When the effects of the poison result in the formation of a toxic metabolite (bioactivation), the therapy is to inject an antagonist that inhibits the

toxin's pathway and speeds up its elimination or excretion from the body. Two illustrations are the treatment of methanol poisoning, which biotransforms to formic acid and formaldehyde in the body, by injection of 4–methylprazole; and the treatment of acetaminophen poisoning (often used in suicide attempts), which causes depletion of glutathione and death of liver tissue, by injection of N-acetylcysteine.

4. *Receptor:* This fourth type of antagonism occurs when two or more agents bind to different sites on the target organ and produce lesser effects because of antagonistic response. An example of a receptor antagonism is the administration of pure oxygen to victims of carbon monoxide poisoning. Carbon monoxide is toxic because it claims the binding sites in the blood necessary for transport of oxygen to the body tissues. Its efficiency at doing this is approximately 210 times greater than that of oxygen. Treatment is by the administration of pure oxygen. A second example is the treatment of organophosphate pesticide poisoning with an injection of atropine. The antagonists for atropine poisoning are physostigmine and diazepam.

Synergism

When two or more chemicals produce a combined effect that is larger than the sum of the individual effects of each chemical $(2 + 2 > 4)$, it is said to be *synergistic*. An example of a synergism is the potentially lethal combination of certain muscle relaxants, like Darvon or Valium, with alcohol. This combination is frequently used by people in suicide attempts.

Potentiation

Potentiation is used to indicate combinations of substances that produce effects far in excess of the effects observed when either substance is taken individually. For example, Antibuse generally has no effect unless the person taking it also consumes an alcoholic drink; in that case the person will become extremely sick and exhibit the symptoms of extreme alcohol poisoning. Acetaminophen by itself has beneficial effects; however, when taken with ethanol it can cause severe liver damage.

TOLERANCE

Tolerance is the ability to endure, without ill effect, unusually large doses of a drug or substance and to exhibit decreasing effects from continued use of the same dose of a drug or substance. Everyone has a tolerance for certain exposure levels but can also acquire or build up additional tolerance after continued use. *Crossed tolerance* is the lessened susceptibility that persons who have acquired a tolerance for one drug or poison may thereafter exhibit toward some other chemical. Drug tolerance is the progressive diminishing of susceptibility to the effects of a drug, resulting from its continued administration.

Two mechanisms appear to be responsible for these types of observed tolerance. The first is the biological alteration of the sites targeted by the drug, substance, or toxin. The second mechanism is a reduction in the quantity of the substance reaching the target site due to improved biological mechanisms, for example, metabolism or excretion. Less is known about the cellular changes in tissue than is known about the dispositional tolerances built up along the pathway to the target organ.

Immunologic tolerance is an immunological response characterized by the development of specific nonreactivity of lymphoid tissues to a particular antigen. Under other conditions the antigen would be capable of inducing humoral or cell-mediated immunity. This tolerance may be a consequence of contact with this particular antigen in fetal or early postnatal life, or it may follow the administration of very high or very low doses of certain antigens to adults. The induction of tolerance to a given antigen does not affect immunological reactions to unrelated antigens. This condition is also sometimes called *immunological paralysis.*

Several other tolerances are related to the immune system. *Adoptive tolerance* is the immunological tolerance induced by the passive transfer from a donor to an irradiated subject of lymphoid cells that were previously rendered tolerant to an antigen. *High-zone tolerance* is an acquired immunological tolerance caused by the presence, in very high concentrations, of an antigen. Under normal circumstances, this antigen would ordinarily stimulate lymphocytes to antibody production at intermediate concentrations. *Low-zone tolerance* is immunological tolerance due to the continuous presence, in extremely small concentrations, (below the threshold required) of antigens. Under normal circumstances, this antigen would ordinarily stimulate lymphocytes to antibody production at intermediate concentrations.

The immune system is discussed in greater detail in Chapter 5.

DOSE-RESPONSE RELATIONSHIP

Another relationship that is fundamental to toxicology is the correlation between the amount of a toxin that is present in the body and its observed effects. To discuss this relationship, we will break it down into the components of (a) making certain assumptions, (b) doing some calculations, and (c) allowing for a margin of safety.

Assumptions

Several assumptions are central to understanding dose-response relationships:

1. The response observed is related to the substance administered.
2. There is a causal relationship between the dose and effect.
   a. There is an agent-specific target or receptor site.
   b. The degree of response is related to concentration.
   c. The concentration at the site is related to the dose.
3. There is a quantifiable method of measuring the toxin.
4. There is an observable and precise means of expressing toxicity.

Calculations

In 1927 Trevan introduced the concept of *lethal dose (LD_{50})*, which is a statistically derived single dosage of a substance that will produce death in 50% of the animals so administered. The LD_{50} dose is usually the first experiment performed with a new chemical. The level below which a substance produces no effect is called the *threshold*. Sometimes this is also referred to as the *no-observable-effects level (NOEL)*. In toxicology we are frequently concerned with the *no-observable-adverse-effects level (NOAEL)*.

If we have a group of animals and administer a poison, some of them will die. If we increase the dose, more will die. For a large enough dose, they all will die. If we plot this relationship on a bar graph, we can observe what dose is required to kill half of the animals. This is the lethal dose for 50% of the animals, LD_{50}.

When toxic gases and airborne toxins are evaluated, the lethal concentration in the air being breathed, rather than lethal dose, is used. We can, similarly, determine a lethal concentration, LC_{50}, for half of the animals, using the same methodology and technique.

Previously, we have discussed undesired or adverse effects. Any of these can also be used in dose-effect relationships. These effects also exhibit their own characteristic thresholds and NOELs. For drugs, and in the medical field, an *effective dose (ED_{50})* is the dose required to induce the efficacious result being sought by the physician. If we use a different end point, that of therapeutic effect, we can establish an effective dose for 50% of the animals.

Another way to present the data is to consider the cumulative effects. That is, we plot the relationship between increasing dose and the percent of the population affected. This generally results in an S-shaped curve if we use the logarithm of the dose on the x-axis. This characteristic curve is the result of interaction between the toxin and the target site of action, that is, the receptor. Technically, this interaction is called the *Law of Mass Action* and is represented mathematically as:

$$A + R = RA$$

where: A is the toxin, substance, or chemical.

R is the receptor site.

RA = % of receptor sites occupied.

If we let   r = number of sites,

then RA/r = maximum of 1

and   E = function of (RA)

where E is the observable effect we discussed previously.

This expression tells us that the observable effect we are measuring depends on the dose that is administered. (See our original assumptions.)

Using a log-of-the-dose scale provides a symmetrical S-shaped response curve (Figure 1–2). The rationale for doing this will become apparent as we discuss concepts such as *potency* and *therapeutic margin*.

Potency is a value that can be used to characterize and rank various toxins by their LD_{50}. A more potent toxin is effective at lower doses. For three toxins, A, B, and C, the most potent is most deadly at the lowest dose, while the least potent requires the largest dose (Figure 1–3).

While the dose-response relationship can be determined for each toxicity of a toxicant, there are certain limitations when using these data. The slopes of the curves are also important in analyzing the effect of toxicants. For example, consider two toxicants with equal LD_{50}s, A and B. A has a steeper slope and at higher doses appears to

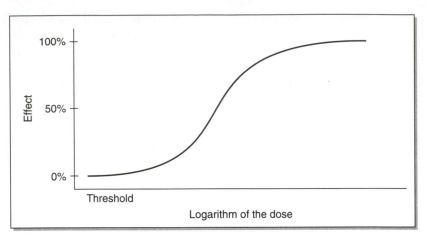

▶ FIGURE 1–2
Relationship between increasing dose and the percent of the population affected.
(Notice that there is a region below threshold where there is a lack of data. Extrapolation
of the curve into this region is problematic.)

be the more toxic chemical, but at lower doses, B has a lower threshold and is more
toxic (Figure 1–4).

Once someone is exposed to a toxicant, the width of the dose-response interval
is as important as the threshold. The difficulty with toxicant A, with a steep dose-
response curve, is that there is much less variability among individuals. Therefore,
once a toxic level is reached, the margin for error of toxicant A is much less than
for substance B. Second, acute toxicity, as generated by lab tests, may not accurately

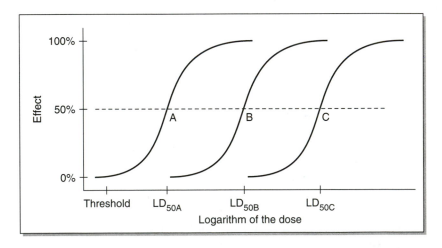

▶ FIGURE 1–3
By plotting the cumulative dose-response curves (log dose), one can identify those
doses of a toxicant or toxicants that affect a given percentage of the exposed population.
A comparison of the values of LD_{50A}, LD_{50B}, and LD_{50C} ranks the toxicants according to
relative potency for the response monitored.

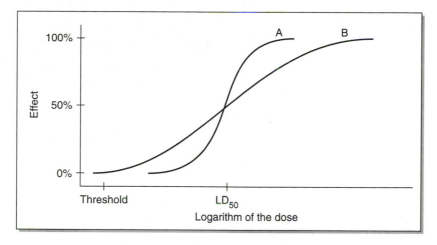

▶ FIGURE 1–4

The shape of the dose-response curve is important. If the LD_{50} values for toxicants A and B were read only from a table, one would erroneously assume that A is (always) more toxic than B; however, the figure demonstrates that this is not true at low doses.

reflect chronic toxicity. For example, both toluene and benzene cause depression of the CNS; for this acute effect, toluene is the more potent of the two. However, benzene is of greater concern to those with chronic, long-term exposures to it because it is carcinogenic while toluene is not.

Margin of Safety

Four terms commonly used in pharmacology and toxicology are the margin of safety, the therapeutic margin, the therapeutic index, and the safety index. The *margin of safety* is the lethal dose 5% (LD_5) minus the effective dose 95% (ED_{95}). The *therapeutic margin* is the lethal dose 50% (LD_{50}) minus the effective dose 50% (ED_{50}). The *therapeutic index* is the lethal dose 50% (LD_{50}) divided by the effective dose 50% (ED_{50}), and the *safety index* is the lethal dose 5% (LD_5) divided by the effective dose 95% (ED_{95}).

The therapeutic margin and the safety margin provide information about the spread of the effective dose and lethal dose curves. Ideally, if there is a choice, it is best to use substances that have wider margins.

The therapeutic index and safety index are numbers. The larger the number, the safer the substance (Figure 1–5).

Other factors such as species, strain, sex, age, diet, and other environmental factors, can affect these values and influence toxicity.

SELECTIVE TOXICITY

Selective toxicity refers to a chemical that produces damage to one kind of living matter without harming some other form of life, even though the two exist intimately. The matter unharmed is called *economic* and that damaged is *uneconomic*.

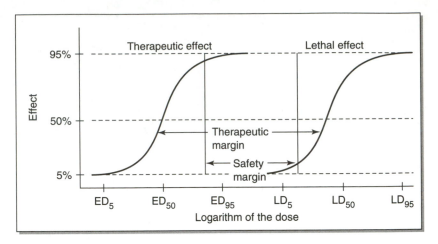

▶ **FIGURE 1–5**
Graphic representation showing the relationship between the therapeutic margin and
safety margin.

Two reasons chemicals may be toxic to one tissue but not to another are (a) the differences in accumulation of the ultimate toxic compound in various tissues and (b)
the ability of their cells to biotransform the chemical. These phenomena allow
chemists to develop selective agents to kill the broadleaf weeds but not the grass.

ANIMAL TOXICITY TESTING

For a variety of reasons, toxicity testing is performed on nonhuman model systems.
Therefore, an important aspect of experimental toxicology is the conversion of animal model system data into terms meaningful to humans. This is done by a process
of scientific extrapolation.

A number of problems arise from extrapolating animal model system data to humans. First, because of the expense of conducting animal studies, the number of animals is minimized. Second, because the normal life span of laboratory model
animal systems commonly used is significantly shorter than that of humans, high
doses over relatively short periods of time are given.

Two principles underlie animal testing. The first is that the results are applicable
to humans on the basis of dose per unit value. The second main principle is that
overexposure of animals to toxic agents is necessary to discover the possible hazards
to humans. Remember, toxicity tests are not designed to demonstrate that a chemical is safe, but rather to characterize what toxic effects a chemical can produce.

Generally, the following information is needed to evaluate animal testing:

▶ Animal species
▶ Route of administration
▶ Time period
▶ Vehicle used to dissolve or suspend material
▶ Dose per unit rate

Common Experimental Protocols

1. *Acute lethality: Acute lethality* is the percent of animals that die in a 14-day period after an injection, unless the likelihood of a substantial exposure may be due to ingestion, inhalation, or absorption by the skin, and then methods are used to emulate such contamination.

2. *Irritation:* Skin and eye irritations are tested in rabbits both dermally and on the membrane of the eye.

3. *Sensitization:* Sensitization, or the potential of a chemical to sensitize skin, is also needed, in addition to irritation testing, for all materials that may come into repeated contact with the skin.

4. *Subacute toxicant:* Subacute toxicant tests are performed to obtain information about the toxicant of a chemical after repeated administrations.

5. *Subchronic tests:* Subchronic tests are given by administering a high, a moderate, and a low dose. Observations of the condition of the animal are made over a 90-day period. At the end of the period, gross and microscopic examinations of the organs and tissue are performed.

6. *Chronic:* Studies of chronic and long-term exposure are similar to subchronic studies except the period of exposure is longer and depends on the intended use in humans. Length of exposure ranges from six months if intended for short periods of exposure to two years if intended for food additive or agricultural implementation.

7. *Teratology:* Teratology and reproduction data are also essential in testing. The embryological development between conception and birth is observed. Information on general fertility is gathered, often for several generations. When developmental malformations (terata) occur, additional studies are conducted to determine the dose necessary (threshold) to induce the malformation.

8. *Mutagenicity: Mutagenicity* is the ability of chemicals to cause changes in the genetic material, DNA and RNA, of the nucleus of cells in ways that can be transmitted during cell division. Mutations in the male's sperm or the female's egg, before fertilization, may result in spontaneous abortion or congenital abnormalities. In vivo (living systems) and in vitro (in glassware or test tubes) methods of fertilization are used for testing for the possibility of multigeneration mutations.

9. *Other Tests:* While most of the tests and observations just described are performed as "standard" toxicant testing protocol, additional tests may be run to provide information about routes of exposure and special effects, such as changes in behavior. These tests would include development of analytic methods to detect residues in tissues, human or otherwise.

PREDICTIVE TOXICOLOGY—RISK ASSESSMENT

Evaluating risk is the main reason for toxicology testing. Toxicant testings and database generation are used to assess the risk (or evaluate the hazard) to humans associated with the chemical. In these tests, the highest administered dose without adverse effects is the no-observable-effects level (NOEL) or no-observable-adverse-effects level (NOAEL). The lowest-observable-effects level (LOEL) is also obtained. To determine the safe dosage for humans, an extrapolation of effects data from high dosages in animal model systems to low dosages in human model systems is performed. Subsequently, it is necessary to relate these data to worker exposure and emergency events.

Safety Factors

Safety factors are commonly used for dosages in humans. A safety factor of ten (10) is typically used for extrapolation from animal to human; another safety factor of ten (10) is used for differences in human susceptibility. This then is a total safety factor of 10×10, or 100 times, for human model system exposure to a toxin. A safety factor of 1,000 is used when no good chronic worker exposure data are available; this procedure is intended to offset the difficulties inherent in extrapolation of animal model system exposure data to workers' on-the-job exposures.

Risk Extrapolation

Risk extrapolation describes the process of evaluating risks associated with exposure to a certain chemical. Almost every aspect of modern living exposes people to health risks. The socially acceptable risk must be determined before threshold limit values (TLVs) are promulgated. The maximum human dosage may be calculated from toxicological testing data. The task of determining safety factors and acceptable risks is a burden that rests on the entire society. It is not reasonable to place this burden solely on scientists and regulators. It is necessary to assess the individual and social cost-benefits of the use of these substances. It is, therefore, important that everyone be aware of the ramifications of the use of hazardous materials in our society.

SUMMARY

▶ Since the early history of humankind, poisons and their effects have been studied. Today this science is well developed into the medical research field of toxicology, a division of pharmacology that studies the adverse effects of chemicals.

▶ The primary focus of this text is on the mechanics of toxicology, which constitute two principal areas: toxicokinetics, or the absorption and movement of substances through the human body; and toxicodynamics, or the observation of the body's response to those substances.

▶ The dose makes the difference.

▶ The hazards associated with a particular substance depend on the toxic triangle: poison, environment, and the organism.

▶ To fully understand the toxicity of exogenous chemicals, or those that are foreign and harmful to the body, research information on cause and effect is required.

▶ Toxicologists need to know where and how a toxic chemical entered the body, what the quality and amount of the dose was, how long and how often the exposure occurred, and what the total chemical burden and personal differences are of the individual subject.

▶ Important components of toxicity include the sites and routes of entry, the concentration or quality of the chemical, the dose or quantity of the chemical that is absorbed, the duration of exposure, the frequency of exposure, and the total body burden.

- The principal factors that affect the concentration of the chemical in the target organs include (a) barriers such as the skin or integumentary system, (b) distribution of the chemical by the rate of blood flow, (c) storage sites for chemicals such as the marrow of the bones, (d) removal of the chemical by the excretory system, and (e) personal differences such as gender, age, and health.
- Undesirable responses of the body to toxic exposure range from allergies to cancer death.
- Chemicals interact with each other to produce a variety of effects from limiting to multiplied.
- An organism can develop tolerance, or the ability to deal with higher levels of exposure to substances.
- The relationship of dose and response is important in determining the effects of a toxin on the body.
- The study of poisons requires researchers to expose a variety of laboratory animal model systems to chemicals. This allows makers of regulations to extrapolate the results of scientific investigations into data that are used to make risk and safety decisions regarding human exposure.

A CLOSER LOOK: DURATION, FREQUENCY, AND DOSE

Let us explore some time-based relationships. First, we need to understand that our body is very dynamic and, as fast as a substance comes in, it is distributed, transformed, and excreted. Therefore, the available, or effective, dose generally varies greatly with time.

When a person is removed from exposure, absorption stops. The retention time of a substance in the body is characterized by its *half-life*, $t_{1/2}$, which is the time it takes one-half of the substance to leave the body by any exit route. This relationship is illustrated graphically in Figure 1–6.

The shape of this curve is classical, an inverse exponential type of curve. In this case, it is very difficult to reach a "zero" concentration, since the best we can do is to remove one-half of the material over

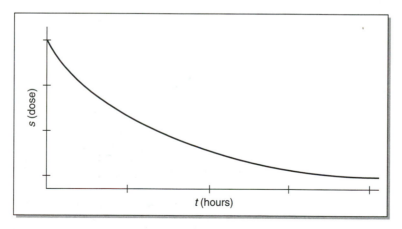

▶ FIGURE 1–6
Decrease in concentration, body burden, in an organism over time. *S* is a measure of the concentration of the "substance." The half-life, $t_{1/2}$, is the time required for one-half of the substance to be removed from the body.

any period of time. An important question, in determining one's level of risk, is, "How long will it take to reduce the concentration below a specified level?"

An important concept is that of threshold. *Threshold* is a measure of going over the edge. Below threshold, nothing happens. "Nothing" is the effect being sought or observed and is the criterion the observer has set. Above threshold, the effect or event occurs. For example, death is an easily measured threshold. Some thresholds, however, are more difficult to determine, for example, being well (subtoxic) and being sick or ill (toxic). At what point, for instance, does a person determine that he is too sick to go to work? Such a point or line is a very subjective determination.

In 1941 the American Conference of Governmental Industrial Hygienists (ACGIH) formed a committee to review available data on toxic compounds. The task of the committee was to establish exposure limits for employees working in the presence of airborne toxic agents. The committee began to publish an annual list of compounds and recommended exposure limits, *threshold limit values (TLVs)*. The primary purpose of the TLVs is to protect healthy male workers in chronic exposure situations.

In the early 1970s, the Occupational Safety and Health Administration (OSHA) was established. OSHA is responsible for adopting and enforcing standards for safe and healthful working conditions for men and women employed in any business engaged in commerce in the United States. OSHA essentially adopted the then-current TLVs, made them official federal standards, and referred to them as *permissible exposure limits (PEL)*. PELs are formally listed in Title 29 CFR, Part 1910, Subpart Z, General Industry Standards for Toxic and Hazardous Substances.

Let us now explore some ramifications of absorbing toxic substances. Assume that there are some limitations in absorption and transport; that is, the actual response of the body to the poison is delayed. Furthermore, let's assume that the absorption is due to a single exposure, an acute event.

When a substance is absorbed, its concentration builds up in the body over time to some maximum level. We will denote this level as S_{max}. After a period of time, this maximum concentration, S_{max}, is reached. The shapes of these dose-response curves are classical, they are typically a variation of an exponential type function.

Now, from our previous discussion of excretion, we know that it takes some time for substances in the body to be removed. Additionally, the removal process follows that of the half-life we previously discussed. If we combine these concepts of delayed absorption and excretion for different doses, we obtain a set of curves as illustrated in Figure 1–7.

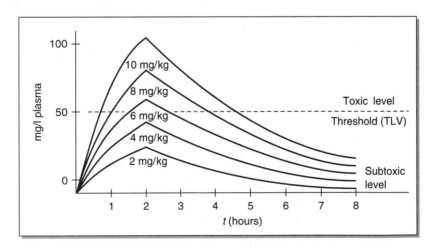

▶ **FIGURE 1–7**
For a single dose, the concentration in an organism will reach some maximum level. The five different curves represent five different doses, expressed, for example, in mg/kg. The threshold limit value (TLV) is a "safe" value. The "body burden" is being measured, for example, as milligrams per liter (mg/l) of blood plasma (blood plasma is the liquid fraction of the blood).

Duration and frequency are important time-based relationships in the study of toxicology. Establishing duration is not always a simple task, but the length of time one is exposed should be obvious to any student who is reading this text: the longer you are in the sun the more likely you are to get burned, and a sunburn is an acute symptom of the skin to exposure from the sun. Too much of anything isn't good for the body. The issues of frequency may sound simple ("Just count how many times you were exposed"), but it may not be so easy. For example, if you work all day in a slightly contaminated area, does that mean one exposure? Maybe, but you took two coffee breaks, a lunch break, and one trip to the restroom, so are these actually five different exposures? These are questions that are typically left for Certified Industrial Hygienists (CIH) to resolve.

No chemical or substance is entirely safe—it is the dose that is important. It is important to quantify both how much of a substance is available at the target site and how long it is there. You should realize that only a small fraction of the substance that you may be exposed to is actually absorbed. Furthermore, only a small fraction of the absorbed dose actually reaches the target site; the rest may be bound, or bioaccumulated. (*Bioaccumulation* is the sequestration of a substance in an organism such that the concentration is greater than might be otherwise expected.)

Now let us combine these dose-response curves to simulate what the "body burden," or actual amount of toxin within the body for a worker over a work week, might look like (Figure 1–8). Let us specify a threshold limit level, the "safe" value or limit. We do not want to exceed this level because we would then experience ill effects. Let us assume exposure and absorption at some fixed rate during the day and excretion based on the half-life.

Another relationship that is fundamental to toxicology is the correlation between the amount of a toxin that is present in the body, how long it remains in the body, and its observed effects. This is known as the *dose-time relationship* and is of tremendous importance in determining whether or not a chemical will be toxic. This relationship is well illustrated by the fact that during the course of a lifetime each of us ingests quantities of many chemicals that, if taken in a single dose, would easily be fatal. An example is that 100 cups of strong coffee contain enough caffeine to kill an average person, yet spread over a long period of time, in smaller doses and greater frequency, this divided dose is just what we may need to get us going in the morning.

A list of foods that have potentially toxic chemicals would fill many pages. Even excessive doses of water, a fluid that is so essential to every bodily function and considered completely nontoxic, can be fatal.

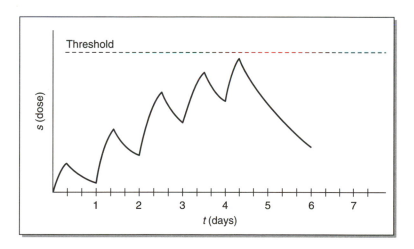

▶ FIGURE 1–8
Situation in which a worker is exposed in the workplace, absorption occurs, and the dose (*S*, or body burden) builds throughout the day. After work, when the worker is no longer exposed, the body burden decreases by normal body excretory mechanisms. The body burden decreases significantly over the weekend.

Conclusion

The relationship between length of exposure and the body's response to an exposure is critical to the student of toxicology. A good illustration of these relationships is the process of getting a suntan. The time of day that we are exposed to the sun is critical because the sun's rays are stronger at noon than at midday or in the morning. The outcome of our tanning session will also be affected by how frequently and how long we come out of the sun (divided dose). Therefore, the length of time that we remain in the sun and how often we go out for an exposure are important elements when considering how not to become sunburned.

Every individual is different and we are all exposed to numerous noxious substances every day. However, the body is also capable of dealing with these exposures if we do not overwhelm our natural defense mechanisms by taking too large a dose at one time or if we give ourselves enough time between doses.

STUDY QUESTIONS

1. What are three common approaches to the subject of toxicology?
2. Briefly discuss each of the following components of toxicity: (a) sites and routes of entry, (b) chemical concentration, (c) dose, (d) duration, and (e) frequency.
3. Name six different factors that affect the total concentration of chemicals in the body.
4. What are the basic personal differences that affect the toxicity of chemicals on the individual?
5. What are the differences between antagonistic effects, additive effects, synergistic effects, and potentiation?

6. What is the meaning of *tolerance* with respect to chemicals that are foreign and harmful to the body?
7. Briefly describe the relationship of dose and response.
8. Name five important protocols for the toxicity testing of laboratory animals.
9. With respect to risk assessment, what are two important considerations?
10. Define, in your own words, the meaning of *toxicity*.

2

Biological Interaction: Absorption to Excretion

COMPETENCY STATEMENTS

This chapter discusses the body's protective barriers to foreign substances; the process whereby chemicals are absorbed, routed, and distributed to the various parts of the body; and the process by which the body removes those chemicals through excretion and other mechanisms. After reading and studying this chapter, students should be able to meet the following objectives:

- Describe the protective barrier known as the epithelium.

- Demonstrate a basic understanding of the various biological mechanisms for the transport of chemicals through or across the epithelium.

- Identify and discuss the different primary routes of chemical absorption into the body.

- Describe how chemicals are distributed throughout the body with focus on (a) volume of distribution throughout the body and (b) localization to the body's various tissues and organs.

- Describe and discuss the various means by which the body rids itself of foreign materials. This will include familiarization with the integumentary, excretory, digestive, and respiratory systems.

INTRODUCTION

Living organisms maintain an internal environment that is different from the external environment; a barrier separates internal from external. In humans, this barrier consists of a special type of cell (the epithelial cell), and anything entering our body (absorption) or leaving (excretion) must pass this barrier. A toxin absorbed into the body is distributed, typically by the circulatory system, through the body. Certain organs, such as the liver, may transform toxins (detoxification) through the process of biotransformation. Toxins interfere with the normal functions of specific "targets" where the material or chemical has its effects. These targets may be cells, organs, subcellular organelles, or specific enzymes. In general, if a toxin can be detoxified or excreted before it significantly adversely affects function, the dose is below threshold and no ill effects occur. Thus, doses below the Threshold Limit Value (TLV) are considered harmless.

The purpose of this chapter is to discuss the body processes that are involved in the absorption, transformation, and excretion of chemicals. These processes are almost exactly the opposite of one another and are essentially the same process the body uses to provide nutrition and remove wastes. The various routes of absorption will also be discussed.

ABSORPTION

The terms *absorption* and *exposure* are frequently confused. *Absorption* is the process by which substances are taken into the body across the cellular membranes that separate the inside of the body from the outside. For example, if you swallow a marble, it will pass out the other end of your gut; it is not absorbed into your body; it does not pass across a cellular membrane. If you take a drink of alcohol, it is absorbed from your stomach into the bloodstream, and you rapidly feel the effect.

Exposure is the condition whereby a substance is presented to the body. When someone is exposed to a substance, it may subsequently be absorbed by and into the body, but no adverse effects can occur until absorption occurs. If there is no exposure, no absorption is possible. The conditions of exposure (such as routes of entry, dose, frequency, and duration) directly affect absorption rates. The conditions of exposure can greatly increase or decrease absorption rates.

The following section focuses on the protective barrier provided by the epithelium and the various mechanisms of chemical transport across or through this barrier. The discussion will include the following mechanisms: diffusion, filtration, permeability, facilitated diffusion, and osmosis, active transport.

Epithelium

The *epithelium* is a layer of cells that forms barriers around and within our bodies. These barriers have specific properties that limit and control which substances may pass across or through the epithelium. Epithelium tissue completely covers the outside of the body—the surface of the skin, the urinary tracts, and the digestive, respiratory, and reproductive systems—all of which communicate with the external environment. Additionally, epithelium also covers many internal compartments such as the brain, eye, inner ear, and joints; cavities such as those surrounding the lungs, heart, and gut; and glands. Because of this arrangement, certain substances

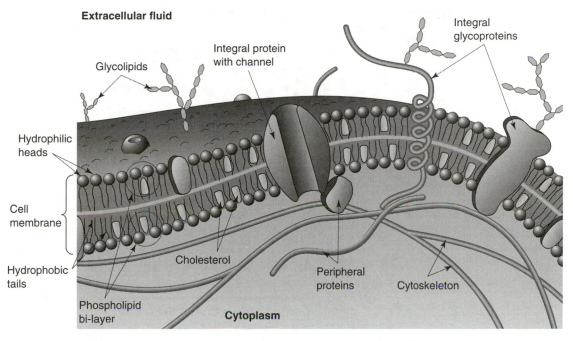

Extracellular fluid

▶ FIGURE 2–1
The cell membrane.

may be sequestered or bioaccumulated in these compartments, a process called *compartmentalization.*

The cell membrane forms the outer boundary of the cell. It is typically a semi-permeable phospholipid bi-layer containing a variety of inclusions such as proteins, and carbohydrates (Figure 2–1). Substances enter the cell by passing across the membrane, a process that may be passive (no work involved) or active (requires energy). Any substance that alters these processes affects function and, therefore, is potentially toxic.

Diffusion

Diffusion is a process in which randomly moving particles scatter themselves throughout the available space. These particles move along a *concentration gradient,* or regions of different concentration. Typically, molecules move from the region of high concentration to low concentration (Figure 2–2). The driving force is kinetic energy. A membrane is a physical barrier to this diffusion; materials will move passively through the membrane if they are small enough to pass through its pores or if they dissolve in lipids (fats and oils) and then pass. Diffusion always refers to particles of matter (molecules).

Permeability

Passive processes of transporting substances across the epithelium depend on the permeability of the membrane. If nothing can pass the barrier, it is *impermeable.* If any substance can cross without difficulty, the membrane is termed *freely permeable.*

▶ **FIGURE 2–2**
Diffusion along a gradient,
high to low.

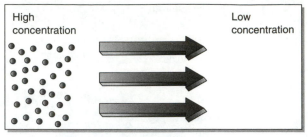

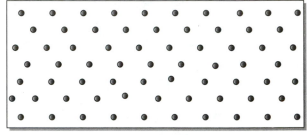

Before: Localized gradient

After: Evenly mixed

Membranes that allow passage of certain substances based on their chemical or physical properties are called *selectively permeable.* Diffusion is the process whereby substances move through a freely permeable membrane because of a concentration gradient, where the concentration is higher on one side (hypertonic side) than the other (hypotonic side). A substance will flow from areas of high concentrations to low concentrations, until the concentration on both sides is equal, isotonic (Figure 2–3).

▶ **FIGURE 2–3**
Diffusion across or through a
permeable membrane.

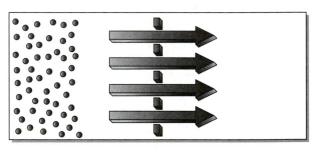

Before

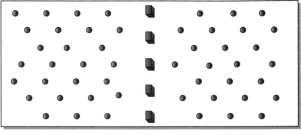

After

▶ **FIGURE 2–4**
Diffusion across a semipermeable membrane.

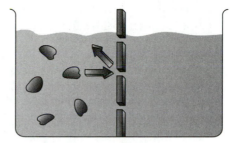

Before: Molecules are too large to pass through holes.

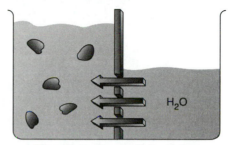

After: Water flows to dilute solution.

Osmosis

Osmosis is the flow of substances across a semipermeable membrane. A typical scenario is one in which the holes in the membrane are large enough to allow a small molecule such as water to cross but too small to allow a large substance, such as starch, to pass. In this case, the water will flow across the membrane to dilute the concentrated solution (Figure 2–4). The increased flow into a cell causes osmotic pressure to build up inside of a cell. In plants, this pressure, called *turgor,* keeps lettuce crisp and leafs flat; it's the opposite of wilting, which is the loss of osmotic pressure. Osmosis always refers to movement of water molecules.

Filtration

During the mechanism of filtration, hydrostatic pressure forces water across a membrane. In these instances, substances (solutes) are selected by size, and smaller substances pass through while larger ones are filtered out (Figure 2–5). The kidney uses this mechanism to filter the formed elements from the blood plasma. Filtration across the walls of small blood vessels pushes water and dissolved nutrients (blood plasma) into the space (interstitial fluid) between cells. This fluid bathes the cells of the body, and substances dissolved in this fluid, such as toxins, can then reach the individual cells of the body.

Facilitated Diffusion

Facilitated diffusion is the process whereby many essential nutrients, such as glucose and amino acids, are passed into the cell. These molecules can be passively transported into the cell across the membrane by special carrier proteins. The molecule to

▶ **FIGURE 2–5**
Filtration due to hydrostatic
pressure or reverse osmosis
(RO).

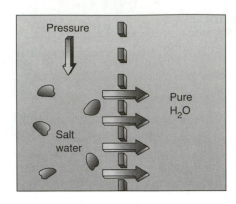

be transported binds to a receptor site on the protein, is moved to the inside of the cell membrane, and is released into the cytoplasm (Figure 2–6). No ATP (Adenosine Triphosphate) is used (no energy spent) in the process, and diffusion is from a high concentration to a lower concentration.

Active Transport

In active transport energy, ATP is needed to move ions or molecules across the membrane. This process is frequently used to create a concentration gradient, for example, sequestering potassium or calcium. The process is complex, and specific enzymes and carrier proteins must be present. Some cells spend up to 40% of their energy on ion pumping against a gradient or exchange pumping, for example, the sodium potassium ATPase pump. This pump sequesters potassium while excreting sodium from within the cell (Figure 2–7).

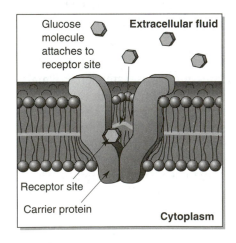

Change in shape
of carrier protein

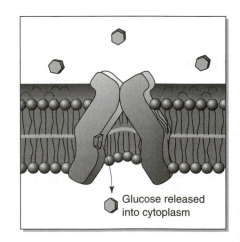

▶ **FIGURE 2–6**
Facilitated diffusion.

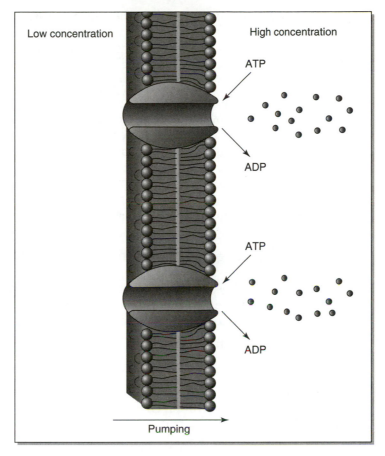

▶ **FIGURE 2–7**
Active transport, Na K ATPase pump, against a gradient.

ROUTES OF ABSORPTION

The process of absorption occurs when substances pass across cellular membranes and enter the bloodstream. The primary routes of absorption in the workplace to be discussed here are:

▶ Inhalation (lung)
▶ Ingestion (GI tract)
▶ Through the skin or integumentary system
▶ Direct injection into the bloodstream.

Eye hazards will be discussed in Chapter 5.

Respiratory System: Inhalation

Most work-related, environmental, and industrial poisonings are caused by contaminated air. The tissues of the lung are exposed to airborne toxins through inhalation.

The depth of penetration depends on the size and weight of the particles. Gaseous substances are absorbed at the level of the alveoli, as is oxygen. Suspension of particles in air also depends on the size of the particles. Insult to the respiratory system may occur in any of the following forms:

1. Dusts or solid particles of relatively large sizes usually deposit in the anterior portion of the nose.
2. Fumes, produced by condensation of molten metals, may, depending on size, deposit along the entire length of the respiratory system.
3. Mists, of suspended fluid droplets, produced by condensation from the gaseous to the liquid phase, may deposit by splashing, foaming, or atomizing of a fluid into a gaseous state.
4. Vapors, converted to this state from either liquids or solids, will deposit by either increasing the temperature or decreasing the pressure.
5. Gases, formless and fluid, occupy space, into which they readily diffuse.

These substances, generally as listed above, are in descending order of molecular weight and particulate size.

The retention of a substance in the lungs is the product of two factors: respiratory volume per minute and difference in concentration in the inhaled and the exhaled air. Therefore, retention is influenced by:

1. Frequency of respiration
2. Depth of respiration
3. Work load
4. Age of the person
5. Temperature of the environment
6. Humidity of the environment

The respiratory tract, which is discussed in depth in Chapter 7 and pictured in Figure 6.1, is equipped with a variety of mechanisms that protect the body from environmental toxins. Large particles are filtered and trapped by coarse hairs in the nose. The bones in the nose are shaped to throw the air outward toward the surfaces covered with mucus. Mucus also traps smaller particles along the respiratory tract. The respiratory tract is lined with cilia, small hair-like processes that beat upward (mucilary escalator) and move the mucus up toward the larynx where it may be swallowed through the esophagus. Minute particles may be suspended in the air in the alveoli and are exhaled during expiration. Simultaneously, absorption may occur throughout this entire pathway, especially by diffusion through the alveolar cell membranes. Some particles that are captured by dust cells remain in the alveolar cells indefinitely.

Gastrointestinal System: Ingestion

The gastrointestinal (GI) tract, the gut, is specifically designed to absorb a large variety of substances. Therefore, it is one of the most effective routes of absorption. Many environmental toxicants are absorbed because they contaminate comested (food), ingested (particulates), or sucked (cigarettes) items placed in the mouth. This is the most common route for the accidental poisoning of children or intentional suicide poisoning of adults. Absorption may occur anywhere along the entire length of the digestive system, from the mouth to the colon (Figure 2–8). As materials pass

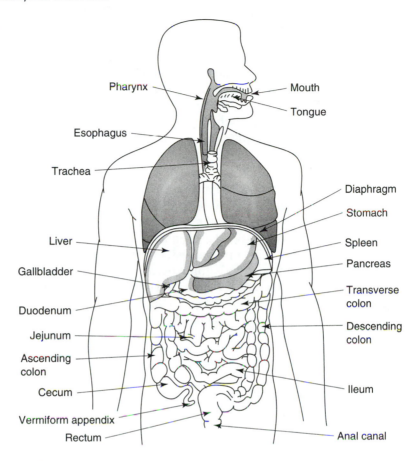

Pharynx

Mouth

Tongue

Esophagus

Trachea

Diaphragm

Stomach

Liver

Spleen

Gallbladder

Pancreas

Transverse colon

Duodenum

Descending colon

Jejunum

Ascending colon

Cecum

Ileum

Vermiform appendix

Rectum

Anal canal

▶ **FIGURE 2–8**
The digestive system.

along the tract, digestive enzymes act on them and various sets of chemical actions and reactions occur.

Absorption from the gut is greatly determined by the pH (acid-base rating) of the contents. The use of neutralizing antacids decreases the absorption of weak acids, but increases the absorption of weak organic bases, such as organic amines. The use of absorbents, such as activated charcoal and kaolin, or precipitating agents, such as milk, may decrease absorption. Activated charcoal and kaolin are often used as antidotes for poisoning. Milk is rich in phosphates, which precipitate the ions of heavy metals, such as lead, and milk proteins precipitate with various caustic substances. Milk cannot, however, in general be used as an antidote or a preventative for poisonous substances.

The major influence on absorption is the rate of motility, or how quickly the substance passes along and through the gut. The longer it takes to pass through the gut, the greater the absorption. Holding a substance under the tongue, for example, allows rapid absorption sublingual (under the tongue) route. This rapid absorption is due to the rich capillary bed in this area and the negligible mobility. Additionally, absorption from the early portion of the GI tract bypasses the liver and its detoxification mechanisms.

In cases in which bioinactivation or bioactivation (see Chapter 3) of substances takes place in the liver, the route of entry of the substances plays an important role. Substances that are rapidly inactivated in the liver are more toxic when they are absorbed after inhalation or through the skin than after oral ingestion. Conversely, substances that are activated in the liver are more toxic after oral ingestion than after other routes of absorption.

Integumentary System: Skin

Human skin comes into contact with, and is therefore exposed to, many toxic agents. One primary function of the integumentary system is to act as a barrier. As such, it prevents a large variety of foreign substances from entering the body and prevents absorption (Figure 2–9). However, many chemicals can be absorbed by the skin in sufficient quantities to produce systemic effects. Routes of entry include the sebaceous (oil) glands, sudoriferous (sweat) glands, and hair follicles. Most chemicals, however, pass through the epidermal cells by passive diffusion. In humans, the permeability of the skin varies depending on the thickness of the skin. The skin is much thicker on the palms of the hands and the soles of the feet than on the broad expanses of the back, buttocks, and thighs. Solvents can facilitate penetration of substances through the skin by providing a medium and by dissolving the substance into a smaller molecule or particulate.

Absorption through the skin is influenced by a number of factors:

1. Moistness caused by perspiration
2. The intensity of the blood supply under the skin
3. Status of the barriers that occur in the skin

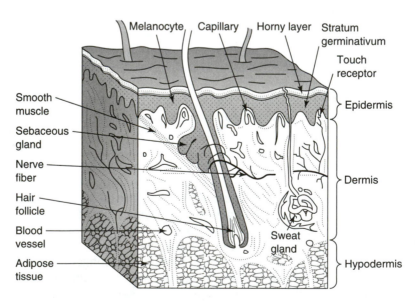

▶ **FIGURE 2–9**
The skin (integumentary system).

These factors are, in turn, controlled by the conditions of temperature, humidity, and activity level.

Injection

Injection is the passage of substances across the skin into the body. Most of you are familiar with the hypodermic needle, a hollow tube that may be fitted to a syringe. This technique is widely used to administer medicine or drugs. Certain reptiles, such as snakes, use this mechanism to deliver venom. Other plants and animals, nettles, and scorpion fish use similar techniques. High pressure liquid may also be injected across the skin. Airless paint sprayers have sufficient pressure to accomplish this. Industrially, leaks from high pressure lines through pinholes may inject a substance into the body. Chemicals can also enter directly into the bloodstream under the layers of our skin when we are cut or scraped by a contaminated object, such as a sharp edge on an old container at a hazardous waste site or stepping on a rusty nail.

DISTRIBUTION

After a toxin enters the body, either by absorption or by injection, it is available for distribution throughout the body. Distribution is highly dependent on whether a substance is water soluble (hydrophilic) or fat soluble (lipophilic). Typically, any substance is partially "bound" and partially "free" (Figure 2–10). The rate at which

▶ **FIGURE 2–10**
Distribution in the body.

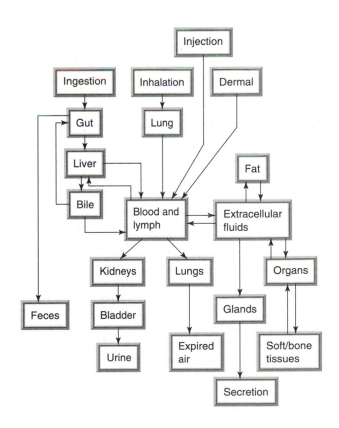

distribution occurs depends on a variety of factors, such as the composition of the blood plasma, the amount of blood flowing to the target tissue, and the cellular uptake. The ease with which transportation and distribution occur depends on the size of the molecule, its solubility in either water or lipids, and whether it is actively or passively transported.

Volume of Distribution

Volume of distribution (Vd) is a measurement of how a substance is distributed within the body. It may be expressed as a ratio between the amount of chemical in the blood as compared to, for example, the rest of the body. If the chemical has a high affinity for a blood component, such as carbon monoxide, which binds to hemoglobin, it may be restricted in a very small volume of distribution. Some chemicals, such as ethanol (grain or drinking alcohol), distribute themselves evenly in the total body fluid (water). Carbon monoxide effects are proportional to the percentage of air composition, whereas ethanol dose-effect is proportional to body weight.

Localization

A toxon is a molecule of a toxin, or toxic substance, that binds to a specific "target" to induce a detrimental effect. If a toxon binds to the receptor site, called the "target site," then this process is inhibited and is an example of a mechanism for a toxic effect.

Toxins concentrate in specific cells, organs, tissues, and spaces. Storage is the sequestering of toxins at sites that are not adversely affected by the presence of the toxin. Bioaccumulation is the localization of toxins at their site of toxic action. Lead and strontium are both drawn to and found in bone. Lead sequestered and stored in bone causes no ill effects. Strontium, however, is radioactive, and when localized in bone, it decays and damages the blood-forming cells of the bone marrow (bioaccumulation).

Blood Plasma Proteins

Plasma proteins can bind many compounds and facilitate distribution of essential nutrients, therapeutic agents, and toxins in the body. Albumins and globulins are the most abundant plasma proteins. Therefore, a significant fraction of most substances are bound to and transported by these proteins. Because of their high molecular weight, these proteins are large. Thus, when toxins are bound to these proteins, they cannot cross cell walls. Consequently, the toxins are "bound" in the plasma and confined to the extracellular spaces and the circulatory system.

Organs

The liver and kidney each have detoxification functions. The liver extracts toxins from the blood that bathes the hepatic, or liver, cells. The binding of proteins and enzymes in the cytoplasm of hepatic cells allows the hepatic cells to dispose of toxins and other substances. Hepatic cells deal with toxins by excreting them into the bile, localizing them, biotransforming them, or some combination of these processes. For example, ethanol (grain alcohol) is sequestered and stored by hepatic cells; an enzyme, alcohol dehydrogenase, converts this into acetaldehyde, which is bioaccumulated and may destroy the hepatic cells, a process called cirrhosis of the liver.

The kidneys function by filtering the plasm from the formed elements of the blood. The filtrate is subsequently scavenged for desirable elements and substances. The remnants, perhaps 1% of the filtrate, become urine and may contain a high concentration of toxins. The bladder, the storage organ for urine, holds this bioaccumulation of toxins, which may detrimentally affect the cells lining the bladder and associated tubes, the ureters and urethra. Additionally, high concentrations of salts can lead to crystallization, for example, kidney stones. Crystallization of oxalic acid can lead to significant organic damage to the functional units of the kidneys, the nephrons.

Adipose Tissue

Fat or adipose tissues are sites in the body that may store high concentrations of lipophilic toxins. Many organic compounds entering the environment are highly lipophilic, which means they readily dissolve in lipids (fats), which commonly comprise 20% to 50% of a person's body weight. Although the fat carried by people may protect them from periods of food shortages, the accumulated fat in their bodies offers storage sites for toxins. Short-term starvation or a sudden increase in mobilization of stored fat for energy (hard work or exercise) can "free" stored toxins that are subsequently distributed to other body tissues. This distribution is a direct consequence of the stored toxins being released from the fatty tissues.

Bone

Bone serves as a reservoir for elements and compounds by incorporating them into its matrix. Such sequestering may or may not be detrimental in situ (in place). For example, although lead exerts a number of toxic effects, it is relatively benign when stored in the matrix. There, it has a relatively long half-life, of about 20 years; thus, it provides low levels of blood lead for extended periods of time. Such low blood lead levels may damage the central nervous system (CNS). Bioaccumulation of some elements is harmful. Fluorine and radioactive strontium have been linked to carcinogenesis. Strontium irradiates the bone marrow hemopoietic (blood-forming) cells, which induces leukemias, for example.

Blood-Brain Barrier

The *blood-brain barrier* is the term used to describe the fact that there is generally not free passage of substances from the blood to the brain. This barrier is composed of glial cells (nurse or helper cells found in the nervous system), called astrocytes, which surround much of the CNS. While less permeable than most other areas of the body, the blood-brain barrier is not an absolute barrier to the passage of a number of potentially toxic substances into the CNS.

However, in contrast to other tissues of the body, the cells of the central nervous system are shielded from both the formed elements (red blood corpuscles and white blood cells) and the plasma of the blood. Before being allowed access to the CNS, the blood plasma is filtered by another glial cell, the ependymal cells, to form cerebrospinal fluid (CSF), which spreads throughout the CNS.

The effectiveness of the blood-brain barrier varies from one area of the brain to another, making it difficult to predict the passage and localization of substances in the CNS. A practical example of the difficulties caused by this barrier is treatment of bacterial infections of the CNS. These infections are notoriously difficult to treat because of the difficulty of ensuring the bioaccumulation of antibiotics at infected areas within the brain.

Placental Barrier

The placental and fetal membranes, amnion and chorion, provide a partial barrier to the unborn child. In the past, it was believed that this barrier protected the fetus from the excesses of the mother. While it is true that free exchange is inhibited, many commonly used drugs, such as tobacco, alcohol, heroin, and cocaine, *can* cross the placenta. In addition to chemicals, many nutrients, viruses (AIDS), and pathogens transverse the placenta. Diffusion is the most common mechanism, and certain factors, for example, lipid-water partition, are important determinants in placental transfer.

Synovial Cavities

The space surrounding our joints, or synovial diarthritic articulations, is a cavity surrounded by a synovial membrane. The membrane filters blood plasma to create synovial fluid. A variety of substances may be so accumulated. Bioaccumulation of uric acid in this space, for example, leads to the formation of uric acid crystals, a medical condition termed gouty arthritis.

Gall Bladder

The gall bladder is a storage organ for bile, a secretion of the liver. While a primary purpose of bile is to solubilize fat to aid in digestion, many hepatic wastes are excreted by means of the bile. High levels of substances can build up while being stored in the gall bladder. High levels of salts crystallize to form bile stones.

EXCRETION

The body has many tissues and organs that are involved in excretion. The principal excretory organs are the kidney, lungs, and digestive systems. The skin and epithelial glands are also means of excretion and include sweat, tears, oil glands, semen, mucus, saliva, and nursing mothers' milk.

Urinary System

The kidney has a unique mechanism to cleanse the blood of impurities. Rather than trying to determine which of, literally, thousands of substances are good and bad, the nephrons simply filter the liquid, plasma portion of the blood from the solid, formed elements. The function of the nephron is then one of reuptake. That is, the cells lining the nephron specialize in transporting specific substances back into the body.

The nephron has three principal regions: the proximal convoluted tubule, the loop of Henle, and the distal convoluted tubule. The cells of the proximal convoluted tubule reuptake substances like glucose, blood sugar, and various plasma proteins. The loop of Henle reuptakes water. The distal convoluted tubules adjust pH and sodium and potassium balances.

Passive diffusion of glomeruli filtrate is a significant reuptake process. When the urine is alkaline (basic), some polar compounds and ions, or weak acids, are not readily resorbed and thus are excreted. The opposite is also true; when the urine is acidic, it more readily excretes alkaline-like chemicals. This knowledge is useful in preventing inadvertent overdosing with certain drugs, such as CNS stimulants (speed) and in treating poisoning by therapeutically introducing the appropriate excretory acidosis or alkalosis.

These mechanisms are also used when biotransformation is needed. For example, organic acids are frequently biotransformed from weaker to stronger acids, thereby increasing the percent in ionic form and thus allowing greater excretion and reduced tubular resorption. There are also two tubular excretory processes, one for anions (negatively charged ions) and one for cations (positively charged ions) in which chemicals are excreted directly into urine by active secretion.

Biliary Excretion

All blood from the gut, below the esophagus, passes through the liver via the hepatic portal circulation system. Therefore, the liver is strategically positioned to sort and review elements and substances absorbed into the blood. Thus, the liver removes compounds from the blood, preventing their distribution to other parts of the body. The liver is the main site of biotransformation of toxins, and the metabolites are frequently excreted directly into the bile. This excretion may present a problem if these substances are then resorbed from the gut, a process known as biliary circulation (Figure 2–11).

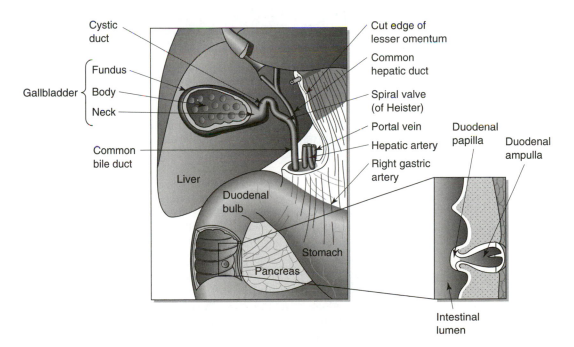

▶ **FIGURE 2–11**
Biliary circulation.

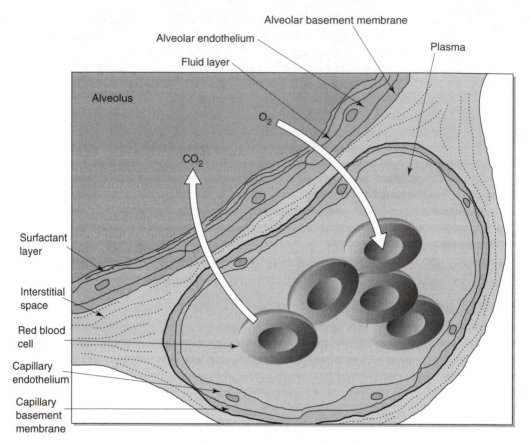

▶ **FIGURE 2–12**
Cross section of alveoli.

Although some of the actual mechanisms of biotransformation remain unclear, many of the effects are quantifiable. It appears that the liver has at least four active transport systems for removal of organic compounds and one for heavy metals.

Lungs

The lungs are the principal organ for the excretion of gaseous wastes; they are also the principal route of absorption of gaseous substances. The lungs remove substances that are predominantly in gaseous form at body temperature. Gases may either be bound, such as oxygen and carbon monoxide on hemoglobin, or dissolved, such as carbon dioxide, in the blood. In the lungs, substances of low molecular weight readily diffuse across the membranes so that the partial pressure of gases in the areolar spaces is essentially equal to that in the blood (Figure 2–12).

Liquids evaporate by producing a vapor. The amount of vapor derived from a specific volume of liquid is related to the substance's vapor pressure. The lungs, thus, also pass liquids as their vapors, for example, water vapor. A practical application of these events is the breathalyzer analysis for drunkenness. The breathalyzer

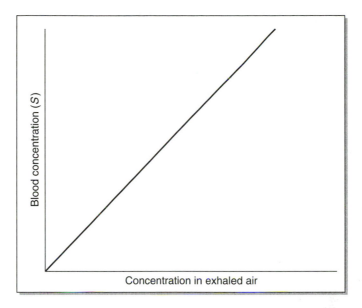

▶ FIGURE 2–13
Relation between concentration, *S*, of alcohol in the blood and the concentration of alcohol in a person's exhaled air.

tests exhaled air to determine the levels of alcohol (ethanol) in the blood. Ethanol is eliminated by a simple system of diffusion. In general, concentration, or partial pressure, of gases or vapors in exhaled air is inversely proportional to blood gas solubility. The less soluble gas is in blood the more will be excreted (see Figure 2–13).

Gastrointestinal Tract

The *gut* is a tube approximately 8.2 meters (27 feet) long that commences at the mouth and terminates at the anus. Substances typically enter the tube at the mouth; nutrients are extracted along the length of the tube; and the remainder is waste and exits as feces. Toxins appearing in the feces may be due to a number of factors:

1. The material was not fully absorbed after ingestion.
2. It was excreted into bile.
3. It was secreted in saliva or the fluids of the stomach, intestine, or pancreas,
4. It was generated by intestinal flora or fauna.
5. It was secreted/caught by the respiratory tract and then swallowed.

The gastrointestinal elimination of lipophilic compounds can be enhanced by increasing the lipid composition of the diet (fats & cholesterols).

Cerebrospinal Fluid (CSF)

CSF of the CNS serves both as a source of nutrients and as a waste excretory mechanism. The same factors that partition off the CNS (blood-brain barrier) also partition substances within the CNS. CSF has a specialized route of removal via the flow of fluid through the arachnoid villi into the superior sagittal sinus of the venus system.

Mammary Glands

Milk serves as a ready storage medium for a number of toxins. Chemicals readily diffuse into milk. Due to its high lipid concentration and slightly greater acidity than blood plasma, many compounds and metals can become concentrated in milk. (These properties explain why the administration of milk with activated carbon is prescribed as an antidote to certain oral poisonings.) While lactation is not generally a significant route for the excretion of toxins, the storage and concentration of toxins by milk pose significant problems for nursing infants. Therefore, diet in nursing mothers requires careful monitoring.

Saliva

Although, on the average, a liter of saliva is secreted daily, this is routinely resorbed from the gut. Unusual substances that are not absorbed from the gut are excreted with the feces.

Sweat

About a liter of sweat is normally excreted per day. Salts and a variety of waste products may be excreted in this manner. Although this is not normally an important route of excretion, the amounts excreted may be significant, which explains why activities that promote perspiration, such as taking a sauna, are sometimes perceived to help detoxify the body. It is possible to sweat out so many electrolytes (salts) that the body becomes hypotonic, subjecting an individual to cramps, miner's cramps, and heat stroke.

SUMMARY

We are all in daily contact with toxic agents (poisons) in the workplace, the home, the garage, on the farm, in the food we eat, the air we breathe, and the water we drink. Fortunately, the body has developed regulatory mechanisms to keep toxins below threshold levels. However, prevention is still the best medicine. Learning to recognize the hazards, prevent exposure, and limit absorption are of primary importance. When the rate of absorption exceeds the rate of elimination, toxins accumulate. When the dose reaches a critical level, the effects become noticeable. Generally, if this dose does not exceed a certain level, called a threshold value, the effects are reversible. The toxicity of a chemical depends on the dose (remember Paracelsus). The effect on the body and its response at any time are determined by a complex set of dynamic relationships including absorption, distribution, biotransformation, and excretion.

In this chapter, we have briefly discussed each of the following relationships:

▸ The epithelium is the layer of cells that forms barriers around and within our bodies, and these cells are found in protective membranes everywhere throughout the body. This includes the surface of the skin, digestive system, respiratory system, and every surface area that communicates with the exterior environment.
▸ Chemicals move across or through the epithelium by several mechanisms, including diffusion, permeability, osmosis, filtration, facilitated diffusion, and active transport.

- Diffusion is the random scattering of particles throughout the available space. Molecules with certain chemical-physical properties may diffuse across cell membranes.
- Permeability refers to the passibility of molecules through the pores of the membrane. If they cannot pass through, then the membrane is considered impermeable. Substances flow through from areas of high concentrations to low concentrations until both sides of the membrane are of equal concentration.
- Osmosis is the flow of water across a semipermeable membrane.
- Filtration occurs when hydrostatic pressure forces water across a membrane.
- Facilitated diffusion is passive transportation of molecules across the membranes and into the cell by special carrier proteins.
- Active transport is when the carrier protein is given a special boost as through activation of the sodium potassium ATPase pump and requires energy.
- The primary routes of absorption include (1) inhalation of air, which is largely a particle filtration system, (2) ingestion of goods and phlegm into the gastrointestinal system, (3) skin or dermal absorption through the oil glands, sweat glands, hair follicles, and passive diffusion through epidermal cells, and (4) injection either under the layers of the skin or directly into the bloodstream.
- Once chemicals have been absorbed by the body, they are distributed throughout the body by the circulatory system in the same manner that it distributes nutrients to the various organs, tissues, and cells. Some areas provide barriers to further distribution, some serve as storage sites for many types of chemical substances, and some bioaccumulate toxons.
- The principal organs of excretory efforts include (1) the urinary system and excretion of urine from the kidneys; (2) biliary excretion from the liver; (3) exhalation of gases from the lungs and respiratory tract; (4) the discharge of feces from intestines and vomit from the esophagus (each a part of the digestive/gastrointestinal tract); and (5) discharge of a variety of other fluids in which the chemicals are dissolved, such as saliva, sweat, tears, and mother's milk.

STUDY QUESTIONS

1. What do you call the body's basic protective barrier and where can it be found in the body?
2. What are the various biological mechanisms for the transport of chemicals through or across the epithelium? Name and provide a brief paragraph about each discussed in the text.
3. Explain what is meant by absorption into the body and discuss the different primary routes of chemicals into the body.
4. How are chemicals distributed throughout the body?
5. Describe and discuss the various means used by the body to remove foreign materials.

3

Metabolism of Toxic Substances

COMPETENCY STATEMENTS

This chapter provides an introduction to the body processes that enable mammals to convert many chemicals to other compounds. This ability, known as biotransformation, allows the body to restrict distribution and enhance excretion. This detoxification process is one of the body's ways of protecting itself from exogenous or otherwise harmful chemical substances. After reading and studying this chapter, students should be able to meet the following objectives:

▶ Define the meaning of biotransformation and describe how it helps to detoxify chemicals that are not natural to the body.

▶ Explain briefly the phase I enzyme reactions.

▶ Explain briefly the phase II enzyme reactions.

▶ Explain briefly the meaning of *extrahepatic biotransformation*.

▶ List and describe the factors that affect the rates of biotransformation.

▶ Define the meaning of *bioactivation* and describe how this phenomenon is dangerous to the body.

▶ Describe the function of the kidney in removing toxic substances from the body.

INTRODUCTION

Students who have not previously taken a physiology course may find the terms in this chapter to be difficult. We urge you to focus on understanding the general concepts of biotransformation and bioactivation and the factors that affect the rates of each, with particular attention to factors of individual variability.

BIOTRANSFORMATION

Biotransformation is the process in which living organisms (hence *bio*) alter or transform exogenous and endogenous chemicals.

All living organisms, including humans, are constantly exposed to foreign substances and poisonous chemicals, or toxins. Toxins may be either of natural origin or synthetic. Because toxins are part of the environment, living organisms have developed mechanisms for removing or detoxifying many of these substances. Toxins that are more water-soluble are apt to be eliminated in the urine, bile, and perspiration or by exhalation. The more lipid-soluble of these chemicals are transported by the circulatory system and are accumulated in the body's lipid depots, in cell membranes and adipose tissue. Therefore, we tend to accumulate the lipid-soluble substances.

Fortunately, animals have developed a number of processes that enable them to chemically convert many lipid-soluble, or lipophilic compounds, into more water-soluble, or hydrophilic compounds. These hydrophilic *metabolites,* or end products, have a reduced ability to diffuse into membranes and be stored in the lipid depots. These transformations, therefore, restrict distribution and enhance excretion. The body's ability to excrete these hydrophilic metabolites by urine, bile, perspiration, and exhalation allows for a pathway for elimination of these substances.

Biotransformation reactions are metabolic processes (occurring within the body) that change the structure and character of a chemical. The toxic effects of many substances are highly dependent on the metabolic fate of the chemical in the body. The body can biotransform a toxic chemical in many ways. A set of processes that biotransform a chemical to an end product is termed a *pathway* and the end product is termed a *metabolite.*

The metabolic pathways we are concerned with are located within our cells, and the individual steps of the pathways are mediated by enzymes. *Enzymes* are specialized proteins in our cells. Some of these can convert a variety of unwanted and foreign chemicals to nontoxic metabolites. Biotransformation processes have been observed in the liver, kidney, lung, intestine, skin, testes, placenta, and adrenal glands. The liver, however, is the principal organ of biotransformation. It contains the greatest variety of pathways, has the largest capacity, and is often referred to as the "detoxification center" of the body.

Generally, biotransformation leads to detoxification. Unfortunately, this is not always the case. Each pathway is a series of steps, between which are formed intermediary products, or "intermediaries." Some intermediate chemicals formed in this process may actually be far more toxic than the original molecule. For example, alcohol (both methanol and ethanol) is initially converted by the body into aldehydes, which are intermediate products that have been associated with blindness (in the case of methanol) and cirrhosis of the liver (in the case of ethanol). Events that produce more active, or lethal, intermediaries are part of the process called *bioactivation.*

Biotransformation may be divided into phase I and phase II reactions. Phase I reactions are the first biotransformation step in what is, often, a multistep process. These pathways lead to the eventual excretion and elimination of the biotransformed products. Phase I reactions are *catabolic reactions,* the process in which living cells break down substances into simpler substances. Such reactions might be expressed as

$$\text{Catabolic} \quad AB \rightarrow A + B$$

Two of these catabolic processes are oxidation-reduction and hydrolysis. Oxidation-reduction is the process by which substances combine with oxygen. Hydrolysis is the process in which water is used to split a substance into smaller particles. Each of these processes essentially leads to detoxification of the original substance.

Phase II reactions are enzymatic processes that use the products of phase I reactions to impart further structural changes. The structural changes are frequently associated with increasing water solubility, which increases excretion and elimination. Phase II reactions are *synthetic reactions,* those in which larger molecules are formed from simpler ones. Such reactions might be expressed as

$$\text{Synthetic} \quad C + D \rightarrow CD$$

In synthetic reactions, an additional molecule, such as a sugar or an amino acid molecule, is *covalently bound* to the parent molecule; that is, a bond is formed when the electrons between atoms are shared. The product so formed is usually more able to be dissolved in water and is therefore termed as *water-soluble. Metabolites* are molecules of substances that have been biotransformed. Biotransformation mechanisms are generally able to produce metabolites that are more water-soluble than the parent molecules and, thus, are easier to excrete and eliminate.

A variety of factors can affect biotransformation. Of particular importance in industrial toxicology is the condition of the liver, nutritional status, and competing processes. The liver is the primary site of biotransformation (which is regarded as a primary hepatic function). Toxins are specifically transported to the liver. Cells of the liver that are so exposed also become one of the first targets for many toxic substances. (The liver is discussed in more detail in Chapter 5.) Conditions that affect the liver, such as cirrhoses of the liver, decrease its efficiency and, thus, its ability to biotransform. Nutritional deficiencies, such as the lack of certain vitamins, coenzymes, and proteins, can reduce the body's ability to synthesize the key enzymes required for biotransformation. Competing processes can be varied; alcohol, nicotine, and other toxicants can directly, or indirectly, affect the biotransformation processes. In some instances, genetic factors may affect biotransformation. For example, some individuals may completely lack the ability to synthesize a particular enzyme that is essential to a specific metabolic pathway. Such an individual may exhibit an "idiosyncratic" reaction to a chemical because it is toxic to them. Accurate prediction of toxic effects is, unfortunately, made more difficult by these individual variabilities.

ENZYME CATALYSIS

A *catalyst* is a chemical substance that, when included as an additional ingredient, facilitates a chemical reaction and causes the reaction to occur at faster rates than would normally occur under a set of conditions of standard temperature, pressure, and concentrations. An *enzyme* is a catalyst formed by living cells to "speed up" chemical reactions with the organism. As a result, alteration of chemicals in the body are mainly enzyme-catalyzed reactions, and there are thousands of enzymes with specific roles in human metabolism. Enzymes facilitate chemical reactions by forming intermediate complexes with reactants that permit structural change of the molecule, which is then released as a product of the transformation. This underlying layer of reactants is known as a *substrate,* and after the molecule has changed, it is the product. The value of enzyme-catalyzed reactions is that they allow reactions to occur under the mild conditions under which we live; for example, our bodies do not need to create a fire to produce heat.

There are also two types of enzymes:

▶ *Simple enzymes* consist of long chains (polypeptides) of amino acids, and they contain only carbon, hydrogen, oxygen, nitrogen, and occasionally sulfur. Simple enzymes are pure proteins such as pepsin and trypsin, which are found in the digestive tract.
▶ *Conjugated enzymes* have both a protein group and a nonprotein group referred to as a *prosthetic group.* This nonprotein group may be an organic molecule called a *coenzyme* or a metal ion such as copper, iron, or magnesium, which is called a *cofactor.* Often both coenzymes and cofactors are necessary for an enzyme to be an active biological catalyst.

The body biologically transforms exogenous, harmful, or foreign chemicals in two major ways. In this chapter we will briefly discuss both phase I and phase II biotransformation.

Phase I Enzyme Reactions

The primary phase I enzyme reactions pathways involve several systems for contacting enzymes to exposed functional groups. These enzymes actually consist of a family of biological molecules that have the capacity to break the bonds between hydrogen and oxygen atoms, carbon and hydrogen atoms, and carbon and oxygen atoms. Enzymes that are responsible for breaking down larger molecules, breaking bonds to add an atom or functional group, or breaking apart bonds between atoms are typically named by adding *-ase* to the end of the function. Thus, a dehydrogen*ase* is an enzyme that removes a hydrogen atom. To *hydrolyze* is to break apart by adding water (H_2O), typically a hydrogen atom to one molecule and a hydroxyl (OH) group to the other. Sucr*ase* is an enzyme that hydrolyzes table sugar (sucrose) to two monosaccharides, glucose and fructose. Water is added (required in the reaction) such that where initially the two molecules were bonded together through an oxygen atom, each molecule eventually ends up with a hydroxyl group at the point where it was previously connected through the oxygen linkage.

Oxidation—Cytochrome P–450

The enzyme complex, cytochrome P–450, is, perhaps, the most important enzyme system involved in phase I reactions. Cytochrome P–450 is actually composed of two enzymes embedded within the cells of the liver tissue. The system contains an enzyme that functions as a monooxygen*ase*. Using this enzyme, the P–450 system transforms many compounds by introducing oxygen atoms into their molecular structure.

The most suitable substrates for oxidation reactions are alcohols, aldehydes, organic acids, simple straight-chain aliphatic compounds, and organic amines. The oxidation of these aliphatic compounds and primary alcohols, aldehydes, and organic acids, is a beta-oxidation, that is, a fragment of two carbon atoms is repetitively cut off and formed as acetic acid. This process is analogous to the oxidative degradation of fatty acids for energy.

The first synthetic detergents (hard detergents) were made from highly branched paraffins, by-products of oil refineries that are unsuitable for motor fuel. These highly branched carbon chains are resistant to the oxidation process. Unlike the classic soaps (sodium and potassium salts of long unbranched fatty acids), which are excellent substrates for various microorganisms, these hard detergents persist as water pollution.

There are a number of other phase I biotransformation oxidative enzymes. For example, amine oxidase oxidizes nitrogen and sulfur atoms.

Hydrolytic Reaction

Ester*ases* and amid*ases* hydrolyze aromatic esters, alcohol, and organophosphates. The actions of most esterases are limited to groups of related esters. Substituents on the carbon atoms directly adjacent to the ester group strongly influence the rate of hydrolysis of the esters. Various esters used as softeners in the manufacture of plastics are highly lipophilic and can diffuse from plastic containers to lipid foodstuffs. The same may occur in plastics used for blood transfusions. Softeners resistant to esterases are undegradable and tend to accumulate in the fatty tissues of the body. Diethylphthalate is an example of a transformation product that can pose such a problem. This product is used in the manufacture of benzoic acid, synthetic indigo dye, and artificial resins (glyptal).

Mammals have high levels of esterases in the plasma and liver. Insects are much less endowed with natural levels of esterases and therefore are much less capable of hydrolyzing esters, alcohols, and organophosphates. Therefore, some organophosphate insecticides contain ester groups, which mammals can hydrolyze and detoxify rapidly. Malathion is an example of an organophosphate insecticide ester.

Several other enzyme systems are also of use. For example, epoxide hydro*lase* catalyzes the hydration of aromatic compounds. Dehydrogen*ases* biotransform the functional groups of alcohols, aldehydes, and ketones by removing a hydrogen atom from, for example, a hydroxyl (OH) functional group and thus permitting their detoxification and removal from the body.

Phase II Enzyme Reactions

The phase II enzyme reaction systems are *biosynthetic* and, by definition, they require energy from the body to drive the reaction. Adenosine triphosphate (ATP) is the "Energy Coin" of the cell. Thus, all energy used by the body's cellular processes is supplied directly or indirectly by ATP. ATP is produced through a process termed *glycolysis,* the breakdown of blood sugar (glucose). This complicated set of reactions occurs partly within the cell and partially within the *mitochondria,* specialized structures in the cell that manufacture ATP. Phase II reactions are accomplished by activating "cofactors" directly or indirectly with ATP.

The following are examples of a variety of phase II reactions.

Glucuronosyltransfer*ases*

Glucuronosyltransferases represent one of the major enzymes that carry out the reactions of exogenous and endogenous compounds to polar, water-soluble compounds. Most of this activity occurs in the liver, but it is also present in the kidney, intestine, skin, brain, and the spleen. Functional groups that undergo conjugation (joining) with glucuronic acid include aliphatic and aromatic alcohols, carboxyl acids, amines, and free sulfhydryl groups. Conjugation with glucuronic acid takes place with alcohols, especially alcohols that are not rapidly oxidized, such as secondary and tertiary alcohols. Glucuronic acid is a relatively strong acid with a large number of alcoholic OH groups and, thus, is very hydrophilic. Furthermore, this group promotes excretion not only because of its increased water solubility, but also because it allows them to participate in the biliary (liver) and renal (kidney) transport systems.

Sulfotransfer*ase*

Sulfotransferase is a group of soluble enzymes whose primary function is to transfer inorganic sulfate to the hydroxyl group present on phenols and aliphatic alcohols. These reactions result in the production of sulfate esters, which are more readily excreted than the original sulfate compound. This detoxification process conjugates phenols, acetacholamines, organic hydroxylamines, hydroxysteroids, and alcohols.

Glycinotransfer*ase*

Typically acids that are not, or cannot, be further oxidized, such as carboxylic acids, form chemical compounds (conjugates) with glycine (a simple amino acid). This type of chemical reaction (conjugation) is especially true for acids with substituents on the carbon atom adjacent to the carboxyl group and for aromatic acids, such as benzoic acid. Benzoic acid is conjugated to hippuric acid and salicylic acid is conjugated to salicyluric acid.

Mercapturic Acid

Organic chlorine-containing (chloro) and bromine-containing (bromo) compounds undergo a reaction in which the chlorine or bromine atom is replaced by a mercapturic (sulfur) acid group. The *mercapturic acid* derivatives are highly hydrophilic and are readily excreted. These substances are good substrates for active transport systems in the kidney and liver.

Methylation

Methylation is a common biochemical reaction of endogenous compounds. Methylation takes place on a single nitrogen atom and on carbon atoms and nitrogen atoms in a ring called a *heterocyclic ring.* Frequently, the quaternary ammonium bases (NH_4) formed are hydrophilic and suitable for active excretion. However, generally, these reactions differ from other conjugation reactions in that it actually masks functional groups that may reduce its water solubility; thus, it may impair the ability to further biotransform the molecule. Nevertheless, the reaction leads to bioinactivation (discussed below) and, therefore, detoxification.

Acetylation

Acetylation occurs between foreign substances and amino groups that are not suitable for oxidative transformation. Aromatic amines, such as aniline, where the amino group is attached directly to the aromatic nucleus, and the alkylamines, where the amino group is attached to a tertiary carbon atom, are candidates for acetylation. The acetylation of the sulfonamides may give rise to certain complications, and undesirable side effects may occur. Fortunately, acetylation generally leads to a decrease in toxicity due to the masking of the amino group, which is an important event for biological activity.

EXTRAHEPATIC BIOTRANSFORMATION

The term *extrahepatic* means that the process occurs in places outside of the liver.

Epithelium

Other cells and organs are important in the regulation of foreign compounds, such as the environmental pollutants found in air, water, and food. The major tissue involved is the epithelial, which is found, for example, in the lungs, kidneys, skin, and the gastrointestinal tract. Epithelial cells are a front line of defense, cover the surfaces of the body, and biotransform many of the prevalent low levels of toxins found in our environment. They can have a marked effect on the ultimate disposition of selected chemicals.

Intestinal Fauna

Intestinal microbial biotransformation is often overlooked. In recent studies, the action of intestinal microbes has the potential for biotransformation as great as or greater than the liver. Unfortunately, these transformations may be either toxification or detoxification reactions.

Chelation

Chelation is a chemical interaction that includes metabolic processes and is an example of an antagonistic effect. Many metals such as arsenic, copper, lead, mercury, nickel, and zinc can be effectively detoxified by formation of a complex metabolite from an organic compound or chelating agent. This metal/chelate can then be excreted from the body by kidney filtration and elimination in the urine.

Therapy for lead poisoning is a good illustration of chelation therapy. A new complex metabolite is formed by ingestion or injection of calcium sodium diaminetetraacetic acid or EDTA and dimercaprol. Mild cases can be treated with N-Acetyl DL-penicillamine. (Such treatments can be life threatening and must be conducted under the direct supervision of a physician!)

FACTORS AFFECTING RATES OF BIOTRANSFORMATION

Intrinsic Factors

A number of intrinsic factors can affect the rates of biotransformation. Examples include the concentration of the compound, other physicochemical properties of the compound (ionic nature), and the dose. The ease with which a compound crosses a cell membrane is governed by its lipid solubility. An important factor in determining solubility is the presence of groups of atoms in the compound that are capable of forming ionic bonds. *Ionic bonds* are formed by complete transfer of an electron from one atom to another and resulting in ions that are oppositely charged and attract one another. Examples of such compounds are the phosphates, sulfates, and phenolic hydroxyls. Depending on the hydrogen potential (pH), these functional groups may be more or less water soluble and, thus, more or less readily transferred across the cell membranes.

Another factor is protein binding. Many substances are "bound" to a variety of plasma, interstitial, or intracellular proteins. Such binding removes the substance from the pool of substrate and effectively prevents any action by or to the substance.

An interesting phenomenon involved in many biochemical reactions is that of stereoselectivity. In simple terms, *stereoselectivity* occurs when a molecule has a pair of isomers (mirror-image molecules that have similar melting and boiling points and solubility, but that are different in terms of configuration or other ways) that are present in the mixture in equal amounts (a racemic mixture). In this situation, metabolism occurs at different rates for each isomer, and while one may appear largely unchanged and still be foreign when found in urine, the other will appear as a fully transformed metabolite.

As previously mentioned, an enzyme reaction has a maximum velocity. As substrate concentration increases, the amount of product formed increases until the maximum is reached. Increasing the substrate concentration further has no effect on the amount turned over (product produced) because the mechanism is saturated (at capacity). Ethanol (grain alcohol) is degraded in a series of steps. The first step produces an aldehyde. If the concentration of alcohol is too high, a pool of aldehyde forms, frequently in the liver. In heavy drinkers, this pool of aldehyde "fixes" the cells of the liver and results in cirrhosis of the liver.

Enzyme Induction

Induction of enzymes, by artificial means, has been demonstrated to enhance the effectiveness of biotransformation. Examples of chemicals that can induce enzymes are prescription drugs, pesticides, industrial chemicals, natural products, and even ethanol. Literally hundreds of materials have been found that induce the enzyme monooxygen*ase*. Such chemical agents have been termed microsomal-inducing agents. The most widely studied inducing agents are phenobarbital and the polycyclic aromatic hydrocarbons. Other major classes are halogenated pesticides, hexachlorobenzene, lindane, polychlorinated biphenyls, steroids, and chlorinated dioxins.

Enzyme Inhibitors

At the molecular level, there are many mechanisms of enzyme inhibition. Generally, these are considered to be toxic mechanisms and will be considered in more detail later. Agents that effect protein synthesis, those that affect the tissue levels of necessary enzyme cofactors, and inhibitors of the P–450 system will all, ultimately, inhibit biotransformation enzymes. These inhibitory actions result in the loss of bioinactivation and biotransformation capacity. When toxins are not enzymatically transformed and eliminated, therefore, the number of adverse effects to the organism increases.

Individual Variability

Variables in the host (individual) include sex, age, genetics, nutritional status, and the effects of environment. The presence of a diseased condition may greatly influence biotransformation.

Sex

Sexual differences in biotransformation are significant. The balance between female estrogen and male testosterone hormones are important in determining the activity of cytochrome P–450 enzymes. Introduction of low levels of testosterone in female rats increases their ability to transform a number of chemicals. Similarly, castration of male rats reduces their capacity to biotransform foreign substances. In humans, sex-dependent differences are observable in the transformation of nicotine, acetylsalicylic acid (aspirin), and heparin.

Age

Both the young and the elderly have decreased ability to handle toxins. Even though not all pathways are absent or limited in newborns, fetal and newborn animals have severely limited ability to biotransform. In the elderly, there is a decrease in cytochrome P–450 and the activity of its associated reductase. These effects are due not only to reduced production of enzymes but also to decreased renal and hepatic blood flows, decreased liver size, decreased efficiency of the excretory system, and increased mass of adipose (fat) tissue.

Nutrition

Diet and nutritional status are important factors in determining toxin transformation and excretion from an organism. Mineral deficiencies decrease oxidation and reduction reactions; fortunately, a shift to a balanced diet and, thus, to a normal nutritional status, usually returns the enzyme activities to normal levels. Deficiencies of vitamins C, E, and B reduce biotransformation. Diets low in protein increase toxicity of a number of foreign substances because of the loss of enzyme production capacity.

Ill Health

Hepatic (liver) injury, and any disease state that interferes with the normal function of the liver, can influence the biotransformation process. Cirrhosis, hepatitis, carcinomas, jaundice, other viral infections, and conditions that decrease hepatic blood flow suppress biotransformation and clearance of toxic substances. Stress negatively affects the immune system, making the body more susceptible to illness and toxins. This may create a rapidly deteriorating situation with a range of diverse consequences.

BIOACTIVATION

Biotransformation processes also have the potential to produce reactive intermediates, or products, with greater toxicity than the parent compounds. The enzymatic formation of these substances is termed *bioactivation*. The factors and variables that are known to affect normal biotransformation also affect bioactivation of toxic substances.

Bioinactivation and bioactivation of substances take place in the liver. Parathion, which itself is not very toxic, is changed in the liver to paraoxon, the actual toxon, which is a very potent, irreversible inhibitor of acetylcholinesterase. Paraoxon is subsequently bioinactivated by hydrolysis into paranitrophenol.

One type of transformation that often leads to reactive chemical species is oxidative dehalogenation. This reaction occurs when the carbon undergoing oxidation contains two halogens (bromides or chlorides). The product of this oxygen insertion is a dihalohydrin, which then undergoes dehydrohalogenation to become a reactive acid called carbonyl halide. A similar reaction is known to occur for chloroform, which produces phosgene, a dangerous chemical that increases the pH of the body fluids and endangers the chemical balance (known as *homeostasis*) of the body.

Often, because of the multiple and parallel nature of enzyme pathways, the product of one reaction leads to the positioning for another enzyme event that may ultimately produce a toxic agent. The biotransformation of acetaminophen is an example of how a minor metabolic pathway can result in tissue injury. Under abnormal conditions this analgesic (pain reducer) can begin a chain of biotransformations that ultimately result in an increased activity of the cytochrome P–450 system and, eventually, to virtual depletion of other needed enzymes. Depletion of P–450 may, under certain circumstances, inhibit other coincidental parallel phase I events. Many other substances also require a similar multistep metabolic sequence that will not occur with the depletion of cytochrome P–450! Therefore, these substances have a much more potent effect, a process known as potentiation.

THE ROLE OF THE KIDNEY

The primary goal of metabolism of hazardous substances is to permit the body to limit their distribution and facilitate their excretion. A primary way of accomplishing this is to make the compound more water-soluble so that it is less likely to cross epithelium membrane barriers into areas of the body where it might interfere with cell functions or cause other internal damage. Water-soluble molecules are also more readily flushed out of the body in the urine.

The role of the kidneys is to filter undesirable substances from the blood. The kidneys are both important and very efficient at this role, especially when the wastes are of high water solubility. Approximately 20% of the blood in the circulatory system passes through the kidneys every minute (about 1 liter, .26 gallon) of blood per minute. The blood passes through the microcirculation of the kidneys, and much of the plasma is filtered into the tubules of the nephron, which is the functional unit of the kidney. Unwanted waste compounds are effectively isolated when the nephron reabsorbs almost 99% of the water and other important minerals and nutrients. The concentrated waste compounds are termed urine and are excreted. Unfortunately, these concentrated toxins in the urine may negatively affect the cells they contact during storage, transport, and micturition (urination).

The vital role of the kidneys as the principal method for removal of toxic substances from the body heightens the concern for those toxins that target the kidneys themselves. The kidneys and the danger and identity of such nephrotoxins will be discussed more in depth in Chapter 5.

SUMMARY

▶ The body transforms certain chemicals into other chemical compounds. When this transformation results in the increased ability of the body to limit distribution and to enhance excretion, it is called biotransformation. When the process produces chemical compounds that are more dangerous and more difficult for the body to handle, it is known as bioactivation.

▶ Phase I enzyme reactions involve several systems for adding enzymes that can break bonds between different pairs of molecules into simpler chemical structures. Possibly the most important enzyme is cytochrome P–450.

▶ Phase II reactions require energy that is provided by the ATP, which is the primary source of energy for the body and is also derived from the breakdown of blood sugar (glucose) in a process termed glycolysis. Many of these reactions lead to creation of new compounds that are water-soluble enzymes that are more readily excreted than the original compounds.

▶ Biotransformation processes can also produce reactive intermediates with greater toxicity than the parent compounds; the enzymatic formation of these substances is called bioactivation, which occurs in the liver.

▶ Reactions such as oxidation, reduction, and hydrolysis essentially lead to detoxification of the original substance and a variety of factors can affect biotransformation. These factors include the health of the liver, general nutrition, lifestyle, eating habits, and natural or genetic differences between individuals.

▶ Finally we learned that the kidney is an important organ in removing hazardous substances from the body through filtration and final excretion in the urine.

STUDY QUESTIONS

1. What is the meaning of *biotransformation* and how does this phenomenon help to detoxify chemicals that are not natural to the body?

2. Where in the body are biotransformation processes observed to take place?

3. In simple terms, what are phase I enzyme reactions?

4. In simple terms, what are phase II enzyme reactions?

5. Briefly describe what is meant by methylation.

6. What is meant by an *extrahepatic* biotransformation? Provide one example.

7. What factors affect biotransformation and the rates of biotransformation?

8. In which organ of the body do the processes of bioinactivation and bioactivation take place?

9. What is the meaning of *bioactivation?* Describe how this phenomenon is dangerous to the body and provide one example from the text.

10. What is the principal filtration organ of the body in removing and excreting altered or biotransformed substances and why?

4

Genetic Toxicology

COMPETENCY STATEMENTS

This chapter introduces the effects of those classes of toxic substances that specifically target the genetic structures. Such toxins result in first-generation somatic mutations of the individual, carcinomas or cancers, and teratogens or mutations resulting in second-generation birth defects. After reading and studying this chapter, students should be able to meet the following objectives:

- Describe the cellular role of nucleic acids and identify the two principal forms.

- Describe the meaning of genetic code and the elements of the code known as "bases" and how "messenger" information is encoded on the DNA and RNA molecules, respectively.

- Define the term *mutagen,* provide a brief summary of their desirable and undesirable effects (mutations), and identify which are natural and which are synthetic.

- List and describe the causes of carcinomas or tumors and explain why some are malignant and others are benign.

- Define the term *carcinogenic* (cancer).

- Identify the different types of cancer and the toxic substances that cause, or are suspected of causing, them.

- Define *teratogen* (terata).

▶ Briefly describe which toxic substances affect the male and female reproductive systems and how.

▶ Describe the government's efforts to protect citizens from exposure to toxic substances and explain why an interdisciplinary approach is needed.

INTRODUCTION

Agents or substances that cause genetic changes are called *mutagens* and the changes, either good or bad, are called *mutations.* When the changes lead to an abnormal growth of tissue cells that are to some extent out of control, it may be classed as a *neoplasm* or *tumor.* Not all tumors are dangerous and may be either benign or malignant. A *benign tumor* is one that does not spread and proliferate throughout the body. A *malignant tumor* is one in which tissue and cells (1) grow more rapidly than normal cells, (2) lose some of the features of the parent cell, (3) invade and destroy adjacent tissues, and (4) spread to other more distant organs. *Cancer* is an umbrella term used to designate a large number of malignant diseases.

A substance that causes cancer is a *carcinogen,* a substance that induces a malignant tumor in humans following exposure. All substances that cause cancer (carcinogens) are mutagens, and most mutagens cause cancer (carcinogenic). When the mutagen affects an egg or a sperm cell, those effects can be passed on to future generations. If the mutagen is able to cross the placenta of a pregnant woman and contaminate the embryo or fetus, it can cause spontaneous abortion and other birth defects. Mutagens that cause such defects are called *teratogens.* This chapter discusses how environmental agents trigger changes in the genetic material (DNA) of living cells.

MUTAGENESIS

As mentioned above, a mutagen is something that causes a mutation, or change, in the original information contained in the gene. This effect may be desirable if it is the basis for an advance in evolution. However, it may also bring about undesirable changes, such as cancer, or cause an undesirable change to occur in some later generation. Figure 4–1 illustrates possible consequences of a mutagenic event.

Often the people actually working with a chemical do not manifest or exhibit any symptom or show any damage, but it is conceivable that damage is being done and could show up some generations later. For mutagenic agents and substances, the time lag between exposure and its effects is usually long term, with mutations not showing up until the next generation or even until several generations later. This long period of latency makes it difficult for toxicologists to identify the event of the exposure and link it to the observable effects.

A chemical substance, mutagen, may either cause changes to the information stored within the chromosomes of a cell or interfere with the operation of the genetic machinery. The following section provides a brief overview of the physiology of a cell.

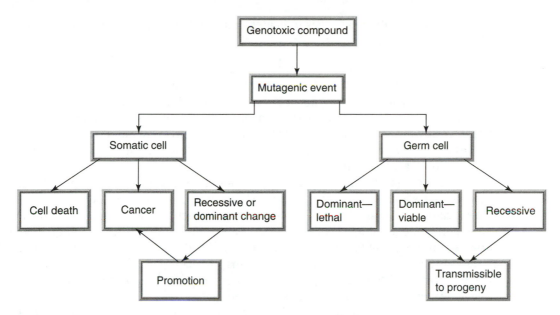

▶ **FIGURE 4–1**
Possible consequences of a mutagenic event in somatic and germinal cells.

The Cell and Genetic Code

The basic biological unit of a living organism is called the *cell.* Our cells contain a nucleus and a variety of *organelles,* each having a specific function.

Within the cell there is also a system of coded messages that tells the cell how to function, to perform all necessary "life" tasks, and to clone itself. This system of coded messages uses an information system known as the *genetic code.*

The cell is managed by molecules known as nucleic acids, which come in two forms: *RNA* or ribonucleic acid, and *DNA* or deoxyribonucleic acid. Information is encoded on the DNA molecule with four main elements or *bases:* adenine (A), guanine (G), thymine (T), and cytosine (C). [RNA uses uracil (U) in place of (T).] These five kinds of DNA and RNA molecules are known as *nucleotides.* The basic building blocks consisting of a sugar (either ribose or deoxyribose), one of the five nitrogen-containing bases, and a phosphate group. In DNA these bases always pair with one another, and their pairing is always in the form of A–T and G–C. A *codon* is a sequence of any three DNA bases and is so named because each sequence codes information and, typically, each code codes for an *amino acid* during the process of protein synthesis. A string of amino acids is a *polypeptide,* and a long strand of polypeptide is called a *protein.* Most functioning proteins are called *enzymes;* enzymes are responsible for performing all of the myriad of tasks required for the upkeep and maintenance of a cell.

The cell creates proteins in the same manner that you and I might write a letter or type out a report: You can think of each code as being a key on the keyboard. We push a set of keys in a certain order to produce a word, a sentence, a paragraph, or a lengthy message. In an analogous manner, the cell creates a message string, which, as mentioned above, causes the formation of a polypeptide by coding a set of amino acids.

Simple probability analysis indicates that four bases, in groups of three, results in 64 possible arrangements:

$$4 \times 4 \times 4 = 64$$

Therefore, we have 64 "keys" on our cellular keyboard for writing a message or genetic code. These many letters or keys can then be linked in different sequences to form many proteins that can perform a large number of tasks. Thus it is with chromosomes, which contain strands of DNA, always occurring in pairs and, when linked together, form a spiral twisted ladder. (A long double strand of DNA, all hooked together, looks a lot like a twisted ladder.)

The "dogma" of molecular biology is "one gene–one protein." When DNA nucleotides are joined together, they form larger units that are called *genes*. The genes are regarded as the smallest unit that can carry a genetic message, and there must be one gene for each single characteristic. A gene consists, therefore, of a sequence of bases comprising a full set of codons. This set of codons, a gene, contains the information required by the cellular mechanisms to create one protein, such as an enzyme. About thirty of these genes have been identified as being involved in the development of cancer and are known as *oncogenes*.

Chromosomes occur in pairs. The numbers and kinds of chromosome pairs, together with their distinctive gene patterns, are referred to as the *genome* of the organism. The chromosomes are what determines whether an organism is an insect, a plant, an animal, or a human. The chromosomes also identify what an organism's physical characteristics are within the species. These physical characteristics can be modified by alterations caused by trauma, the environment, or other factors, but these modifications are only in the organism and are not passed from parent to child. Physical features that are inherited are those dictated by the genetic code obtained from the parents. Figure 4–2 illustrates the structure of the DNA unit and a DNA molecule.

Note: the number of chromosome pairs differs with the species, and all members within the same species have the same number of chromosomes: a fruit fly has 4 pairs and a human has 23 pairs.

Mutations

When mutagens alter the structures of either DNA or RNA, they play havoc with the cell's functions: a mutation has occurred. A change in sequence of the nucleic acids changes the genetic code of the cell itself and thus results in a change in the cell's message: now the cell doesn't know what it is supposed to do!

Both physical and chemical mutagens occur naturally in the environment. Many people assume that mutations are always harmful. This is not the case, as some (very few) are actually beneficial. Darwin's theory of evolution depended on mutations to establish new traits in a population. Beneficial mutations may allow an individual a better ability to run or fly away from their enemies, or in some other way allow an organism to develop better skills of survival. Coupled with Darwin's theory of natural selection, or survival of the fittest, we have a plausible explanation of how life has evolved to the point where we observe it today.

When a cell is mutated, if it still has the ability to reproduce itself, it will divide into two identical cells and they will carry and reproduce the new code. The two new cells, and their progeny, will both be different from the parent cells. These differences

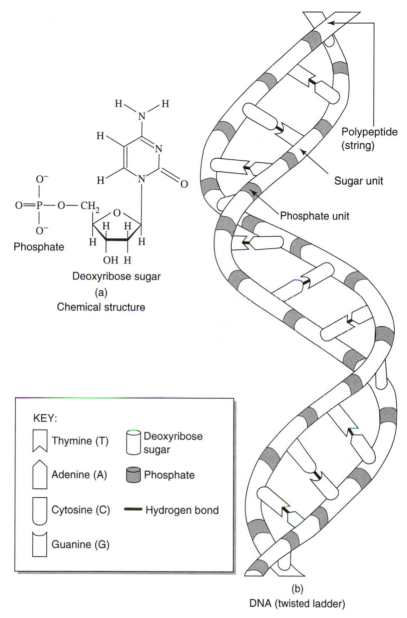

H—N—H

O⁻
|
O=P—O—CH₂
|
O⁻

Phosphate

Deoxyribose sugar

(a)
Chemical structure

KEY:

Thymine (T)

Adenine (A)

Cytosine (C)

Guanine (G)

Deoxyribose sugar

Phosphate

— Hydrogen bond

Polypeptide (string)

Sugar unit

Phosphate unit

(b)
DNA (twisted ladder)

▶ **FIGURE 4–2**
Structure of DNA.

may be so small that the new cells seem to be identical in form or function, or the differences may be very noticeable. See Figure 4–3.

Many mutagens occur naturally. For example, radiation has the potential to damage biological molecules and cause genetic changes. Whether the source is from the sun, from outer space (cosmic), or from soil and rocks, radiation is present in our world, and every living thing is exposed to natural background levels. The

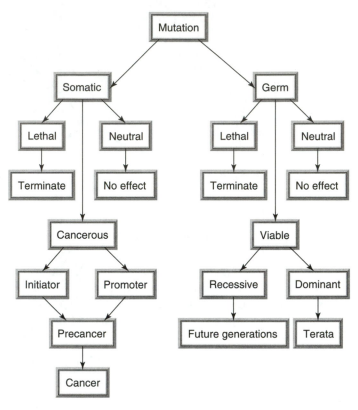

▶ FIGURE 4–3
Mutagenic event ramifications.

atmosphere, the ozone layer, and the magnetic shield surrounding the earth protect us and mitigate these dangers.

Another major natural mutagenic factor is benzopyrene. This mutagen, which is known to cause cancer, is found everywhere in our environment as it is the by-product of the burning or cooking of any organic material. When we make a fire and cook our food to destroy the harmful bacteria we coincidentally also produce benzopyrene. Thus the phenomenon lends credibility to the argument that we, as humans, should eat as many as possible of our fruits and vegetables uncooked and avoid or limit the practice of barbecuing.

While mutations play a major role in bringing about the tremendous diversity of human characteristics, they also cause much suffering and misery: they may also cause cancer. Mutations are, thus, a two-edged sword, and while many may be essential to the ongoing evolution of the species, the majority are harmful.

Three general types of events may cause changes in DNA and RNA, the basic nucleic acids: infidelity in DNA replication, disruption of cellular function, and chromosomal aberrations.

1. *Infidelity in DNA replication:* The enzymes involved in biotransformation may fail to "read" the message of the DNA, which results in passage of a message that is not "faithful" and thus is bad information.

There may also be an interruption of the transcription process, resulting in missing "chunks" of information. This would be like typing out a message on the keyboard, but either the person to whom you are writing cannot read or you ran out of toner or paper and were unable to complete the message in its entirety.

2. *Disruption of cellular function:* Corruption of a single codon due to the insertion or deletion of one or more base pairs from the linear sequence of a DNA molecule can lead to disruption of cellular function. The result would be like typing a message when some of the keys on the keyboard have been traded; when you expect to hit a consonant, you get a vowel or the wrong consonant instead. The result is a message that may not be possible to decipher.

3. *Chromosomal aberrations:* Breaks or gaps in any one complete copy of a DNA strand (chromatid) may lead to the dire consequences of embryotoxicity, congenital malformations, growth retardation, or mental retardation. This is similar to using a keyboard with some of the keys either missing or not working at all.

Biochemical Repair Mechanisms

It is fitting to end this section on mutagenesis by commenting on the human organism's incredible ability to develop a number of repair mechanisms that protect against mutations. One of the most significant is the method of reading the DNA code. When it recognizes that something is wrong—a message is either changed or mistaken—the body corrects the message and reads it the way it should be written. This is much like a built-in "spell-corrector" on a word processor. Persons with fewer built in "mutation-correctors" are predisposed to a larger number of cancers.

CARCINOGENESIS

Carcinogens are substances that cause cancer and induce malignant tumors in humans following a reasonable exposure. According to the National Institute for Occupational Safety & Health (NIOSH), a substance is considered a "suspected carcinogen" if it produces cancers in two or more animal species.

It is conservatively estimated that approximately one American in four will suffer some type of cancer during his or her lifetime. Many diseases are categorized as cancer and, taken together, the incidence (number of cases) should be viewed as being of epidemic proportions in the United States.

The *primary factors* that have been related to the incidence of cancer are:

◗ The genetic pattern of the host or individual natural vulnerability
◗ Radiation, including sunshine
◗ Hormone imbalance
◗ Diet and elements of our lifestyle
◗ Exposure to certain viruses and chemicals

Although it is known that many chemicals do induce cancer and many others are suspected, we still do not understood fully by what mechanism it occurs. It is not really known why some chemicals cause cancer and why others do not. Regardless, it is well documented that some chemicals are carcinogenic and in many respects are similar to other drugs and toxic agents. Carcinogens are *similar* in that they:

◗ Show dose-response relationships

▶ Undergo biotransformation
▶ Vary with sex and age
▶ Interact with other environmental agents
▶ May have their effects either enhanced or decreased

In some respects, carcinogens may be different from other toxins, for example, whether carcinogens exhibit the phenomenon of "threshold" is a point currently under much debate in the scientific community. Carcinogens are *different* from other drugs and toxic agents in that they;

▶ Have cumulative, delayed, and yet persistent biological effects
▶ Demonstrate over time that divided doses may be more effective than individual large doses
▶ Interact with DNA and other molecules with a distinct underlying mechanism

The term *carcinogen* literally means giving rise to carcinomas such as epithelial malignancies. Increases in the types, frequency, early development, and multiplicity of tumors are criteria used in assessing carcinogenicity. When these effects are produced by chemicals or substances, these substances earn the designation of carcinogen. Substances producing any of these effects comprise a very diverse collection of organic, inorganic, solid state materials, hormones, immunosuppressants, promoters, and a number of unclassified carcinogens (Figure 8–6).

Today, cancer is second to heart disease as a cause of death in the United States. Of the cancers involving major organs, lung carcinoma is the most common malignant tumor in men; breast carcinoma is the most frequent in women.

It is likely that cancer is the most dreaded disease of all and, although many other diseases are as destructive to the quality of life and threatening to its very existence, it is probably the rare person who does not give a sign of relief when he is told he doesn't have cancer but has some other disease that may be just as debilitating or deadly. The public seems to have relatively little concern for chemicals that destroy the kidneys, render the lungs nearly useless, or devastate the heart and nervous system, but chemicals that are suspected, or known, carcinogens are set apart from others and have become the focus of special regulations and controls.

Cancer is a process by which cells undergo distinct molecular changes, rendering them abnormal. During the stages of uncontrolled growth, or malignancy, the abnormal cells spread by invading, dividing, growing, displacing, and frequently, destroying normal cells. When cells duplicate themselves without control, the excess cellular mass is known as a tumor, or neoplasm. *Benign tumors* are composed of cells that are not rapidly dividing and are not spreading to other parts of the body. *Malignant cells* grow and divide rapidly.

As these cells invade the surrounding normal tissues, they compete with normal cells for nutrients and space. This leads to eventual atrophy (cell mass loss) by the death of the cells of the healthy tissue. Interestingly, patients are often not killed by the primary tumor but, rather, by secondary infections or as a result of *metastasis* (spreading) to, and destruction of, other parts of the body. Cells of metastasized tumors are difficult to isolate and destroy. Often the cells may completely detach from the primary tumor, metastasize, and invade the body cavity or enter general circulation through the blood or lymph. Once diffused through the circulatory systems of the body, the malignant cells can invade virtually any organ of the body. Once so dispersed, they can develop into secondary tumors. When the growing tumor compresses a nerve or causes inflammation that blocks ducts or circulation, the result is intense pain. The pain associated with cancer is frequently extreme and severe.

Common Characteristics

Cancer is a process by which cells undergo some basic change that allows them to grow without limit and without contact inhibition. Malignant cancers exhibit the characteristics of:

1. *Cell proliferation:* Growth much more rapid than that of their normal parent cells
2. *Loss of differentiation:* Reversion to more primitive cell line status with the loss of some of the features typical of their parent cells
3. *Metastasis:* Invasion and destruction of adjacent tissues and the spread to distant locations where they establish secondary cancer foci

Benign cancers, or tumors that are not malignant, have not metastasized. Whether or not benign tumors have the potential to become malignant is the subject of much debate. The early stages of both malignant and benign tumors are similar. It is hypothesized that, if an organism lived long enough, benign tumors would eventually become malignant. Regulatory agencies have resolved the issue by classifying all chemicals that produce neoplasm, benign or malignant, as carcinogens. Table 4–1 compares the two types of cancer: malignant and benign.

Stages, or Periods of Cancer

Cancer is a multifaceted process that may evolve in any number of ways, as demonstrated by the flow chart in Figure 4–3. It is hypothesized that chemical-induced carcinogenesis involves the following stages:

Preinduction Period

This period is the length of time that, if the organism is removed from exposure during this period, the subject will not develop cancer. This length of time decreases with increasing doses and time intervals. This means that a brief, one-time exposure does not spell doom for the exposed person: people who are accidentally exposed to

▶ TABLE 4–1
A comparison of malignant and benign tumor characteristics.

Malignant Tumors	Benign Tumors
Infiltrates surrounding tissues	Grows by expansion and pushing aside surrounding tissues
Nonencapsulated	
May reoccur after removal of malignant tumor	Often encapsulated
	Does not reoccur
Hardly resembles tissue of origin	Resembles original host tissue
Call division is extremely rapid	Cell division is relatively slow
Tissue destruction is extensive	Causes little tissue destruction
Spreads to different parts of the body	Does not spread
Always lethal if not treated	Usually not lethal, but may become lethal

low levels of chemicals suspected of being carcinogens will probably not develop cancer. The *initiation stage* is the biomolecular event of the start (initiation) of the development of cancer. When a person in an occupational setting is exposed day after day over many years, the question of when the cancer began to grow becomes obscured: Did the cancer begin on day 1, or just how many years later?

Induction or Latent Period

This is the length of time between the initiation of a cancer by a carcinogen and the clinical appearance of symptoms indicating a cancerous growth. This period is also often described as the second stage or *promotional stage,* in which physiological and biochemical changes facilitate the growth and expression of the initiated cell. Molecular events during this period are not fully understood, and it should be realized that a diagnosis of cancer, due to exposure that initiated a cancerous growth, may not be made until many years after cessation of the exposure. The induction period for a given carcinogen varies with the dose: the greater the dose, the shorter the time between exposure and manifestation of the disease. The current understanding of the mechanism(s) of cancer is limited; it is presently undergoing significant evolution.

Genotoxic and Epigenetic Carcinogens

Historically, carcinogens have been divided into genotoxic and epigenetic groups, and each is briefly discussed in the following paragraphs.

Genotoxic Carcinogens

Genotoxic carcinogens are substances that are primary, direct acting, and interact with or alter DNA. These substances result in the conversion of the cell to the preneoplastic stage where they become known as precursor cells. These carcinogens are known as *initiators,* and some examples of such direct-acting primary carcinogens include the following:

▶ Synthetic plastics
▶ Benzopyrene (an active component of coal tar)
▶ Mustard gas
▶ N-dimethylnitrosamine (NDMA), a constituent of cigarette smoke
▶ The nitrosamines
▶ The inorganic metals: cadmium, chromium, and nickel

Epigenetic Carcinogens

Epigenetic carcinogens is the name given to the broad group of agents that promote the growth of preneoplastic cells within the target tissue without evidence of interaction directly with the genetic material. They are thus known as *promoters.* Examples of some of these promoters include the following:

▶ Asbestos
▶ The estrogens and androgens
▶ Azathioprine
▶ Ethanol
▶ Many organic solvents
▶ Tetrachloroethylene

Types of Cancer

In cancer nomenclature, there are four basic types of cancers that are each closely related to the different types of cells or tissue lines that they affect:

1. *Leukemias* are cancers of certain white blood cells and the bone marrow from which they are derived.
2. *Lymphomas* are cancers affecting the tissues of the lymphatic system. An example of a lymphomatic disease is Hodgkin's disease.
3. *Sarcomas* are cancers of the connective tissues of the skeletal system such as bones and cartilage.
4. *Carcinomas* are the most common type of cancers and affect the epithelial tissues that line the protective surfaces, both on the inside and outside, of our bodies. Examples of carcinomas are the cancers of the skin, the linings of the lung, throat, and stomach, and cancers of the excretory tract such as the colon.

Causes of Cancer

In recent years and, for that matter, over the past several decades in a fight against cancer, much research has been done in searching for cures for cancer. Finding the direct connections to the causes has been elusive. Earlier in this chapter, we provided a list of the primary factors related to the incidence of cancer. The following paragraphs provide further, although brief, discussion and examples of several of the factors. In a few cases, a more definite connection has been established, although a clear cause-and-effect relationship may not yet have been identified.

Genetics of the Host

Those involved in the research of cancer and its causes agree that certain people are born with a propensity to contract cancer. As discussed earlier, the body itself has defense mechanisms to try and intercept and reinterpret miscoded messages in the DNA. Quite possibly these are capabilities that many people are born with while many others are not. The theory that a person may inherit this susceptibility suggests the possibility of breeding susceptibility from a population.

Environmental Factors or Geographical Differences

Persons who have made epidemiological studies of cancer have hypothesized, and this is backed up by their data, that geographical differences are linked to the incidence of various types of cancer. For example, it is known that in only a few generations, the offspring of persons who have migrated to a different community begin to exhibit incidences of cancer that are typical of their adoptive community. As early as the 1950s, data from scientific studies suggested that a high percentage of all cancers was caused by "environmental" factors. By the 1960s, many of these factors became linked to certain *environmental chemicals*. Through the 1970s and into the 1990s, it was determined that the air, food, water, and background levels of the environmental pollution affect local communities and contribute to the incidence of cancer in these different population groups.

Lifestyle of the Individual

Some studies have indicated that a high percentage of cancers can be attributed to lifestyle, diet, and personal behavior. These factors include the use of tobacco and may include other habit-forming products such as those containing alcohol and caffeine. The association of colon cancer with the lack of dietary fiber and the association of early sexual behavior with cancer of the uterine cervix are examples. OSHA/NIOSH and other researchers now have an abundance of evidence that secondhand smoke causes cancer. Additionally, many toxins contained in the tobacco smoke have negative synergistic effects when combined with other environmental contamination. For example, an asbestos worker who also uses cigarettes is 10 times as likely to contract lung cancer as a nonsmoking asbestos worker, and 50 times more likely as a normal nonsmoking non-asbestos worker.

Good nutrition includes eating foods that are high in vitamins, like A and C; and trace elements, such as zinc and selenium, can help to strengthen the immune system and, therefore, fight off cancers. On the other hand, obesity, stress, hypertension, cured foods, red meat, and fats lead to higher risk of breast, colon, prostate, and stomach cancers.

Viral Origins

The theory that a fraction of cancers is due to viral infections has yet to be disproved. However, although no anticancer vaccine has been found or developed, there is a growing body of evidence from scientific research that indicates viruses are involved in the development of some cancer. Of course, it is hoped that, eventually, vaccines will be developed to immunize people against these cancers.

Radiation

It seems that every time we pick up a magazine these days, there is an article warning people of the dangers of overexposure to the sun's rays and to the risk of getting a "fake bake" in the neighborhood tanning salon. In an article published in *Science* magazine in 1979, ("Research news: Cancer and the environment: Higginson speaks out"), Dr. John Higginson is reported to consider sunlight (ultraviolet rays) to be responsible for approximately 10% of all skin cancers. More recently, the role of cosmic and natural radiation has been studied, and it has been confirmed that we are constantly being bombarded with high-energy radiations from a combination of sources, such as radon in the rocks and soil and cosmic rays from outer space. Knowing this makes it more difficult for epidemiologists to establish direct cause-and-effect relationships between the incidence of cancer and exposure to trace amounts of synthetic chemicals. Chapter 8 contains a more in-depth study of the health effects of radiation.

Synthetic Chemicals

The net effect of exposure to synthetic chemical carcinogens is very harmful. Various petroleum products and coal tar cause skin cancer, and asphalt and paraffin oils are clearly implicated as potential subcutaneous carcinogens. Inorganic salts, such as those of chromium, are associated with lung cancer, and disorders of the blood leading to leukemia are associated with exposures to benzene.

Additive Effects and Threshold

The existence of thresholds for chemical carcinogens is an area of excited scientific debate. The existence of thresholds for chemical carcinogens continues to be rejected because their existence cannot be experimentally verified. Part of this is due to the inability to accurately extrapolate dose-effect curves to low and near zero doses. The existence of many natural mutagenic agents and events, as well as the pervasiveness of "environmental" chemicals, creates difficulties in establishing true control groups.

Additive effects imply that exposure to chemical carcinogens are additive over the lifetime of the organism. Carcinogenic substances are subject to the same bioaccumulation, transformation, and excretion principles previously discussed. However, if no threshold exists, and a risk is assumed for any absorbed dose, then the probability of cancer induction would be additive (Figure 4–4.)

There are practical limits that may be used for all carcinogens. For example, if exposure to a carcinogen is small enough to reduce cancer incidence to one in a trillion, the level of exposure is of no practical significance, since there are not that many people in the world. If the dose can be related to induction period, that is, larger doses lead to shorter induction periods, decreasing the dose to induction periods greater than 100 years extends the induction past the lifespan.

Although, certainly, not all lung cancer is a result of smoking, experts estimate that up to 90% of lung cancer deaths are directly attributable to smoking. Generally, smokers have a 10 times greater incidence of cancer and a greater risk of dying from lung cancer than do nonsmokers. In addition, as mentioned previously, smoking and occupational exposure to carcinogens can have a synergistic interaction.

Regardless of whether the highest incidence of cancer occurs from one factor or from a combination of all of these factors, it should be clear that the people at greatest risk of developing cancer from exposure to chemical carcinogens are those who are exposed to the highest concentrations of these agents. The workplace contains the greatest and most varied exposure to chemicals, and there are distinct patterns of cancer incidence among workers in certain occupational groups. Industrial pollution of our air, our water, and our soil cannot be tolerated and should not be sanctioned.

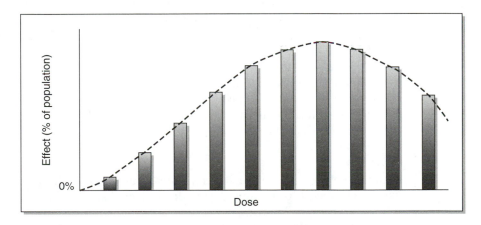

▶ FIGURE 4–4
Bar graph illustrating the relationship between increasing dose and the percent of the population affected. The dotted line represents a curve fitted to the data.

Target Organs and Organ Systems

For the purposes of industrial toxicology, it is advantageous to look at cancers associated with particular organs or organ systems. The following paragraphs briefly describe these target organs. (These organs will be discussed more in depth in later chapters.)

Lung Cancer

Lung cancers are technically bronchogenic carcinomas and pulmonary adenocarcinomas. The smoking of tobacco is considered to be the primary cause of bronchogenic carcinomas. Metal fumes of cadmium, nickel, and chromium are also cancer promoters. Pneumoconiosis, silicosis, asbestosis, and anthoconiosis are also known promoters of cancer.

Urinary Bladder Cancer

Cancers of the bladder are typically epithelial carcinomas. Prolonged contact between the concentrated toxins and the transitional epithelium occurs during urine storage. Polycyclic aromatic hydrocarbons and the active metabolites of benzidine are stored in the bladder.

Liver Cancer

Hepatocellular carcinomas are characterized by hepatomegaly with multiple scattered nodules and large malignant tumors throughout the liver. Many nontoxic chemicals, procarcinogens, are metabolized to a reactive carcinogenic chemical. An example of this is benzidine, which is metabolized to an active form in the liver. Polycyclic aromatic hydrocarbons have also been associated with liver carcinomas. Cancers of the biliary tract are called cholangiocarcinomas.

Leukemia

Leukemias are frequently caused by chemically induced bone marrow changes. Chemical inducers of leukemia include benzene, chloramphenicol, and phenylbutazone and it is also linked to radiation exposure.

Skin Cancer

Epidermal carcinomas are of various types. Squamous cell carcinomas are commonly associated with exposure to ultraviolet light. Contact with petroleum products may cause promotion, and photoactivation may occur. Tobacco products can promote carcinomas within the mouth. Polycyclic aromatic hydrocarbons have been associated with scrotal cancers.

Breast Cancer

About one of every 10 women develops breast cancer in her lifetime. Women who are overweight, have high fat intakes, and drink alcohol are at greatest risk.

Colon and Rectal Cancer

Other than skin and lung cancer, cancers of the colon and rectum occur more frequently than any other. The final opportunity for the body to absorb water is in the last few inches of the gastrointestinal tract. Unfortunately it is also a place where poisons are sequestered, which increases the exposure to adjacent tissues. Therefore, high fiber and low fat diets improve health and protect against toxins.

Oral Cancers

Oral cancer is directly linked to the use of tobacco products and excessive drinking. It can affect any part of the gums, lips, tongue, and mouth or throat.

Cancer Therapy

Treatment of cancer is very difficult and is a special branch of medicine known as *oncology*. The basic challenge of the *oncologist*, or cancer specialist, is the need to rid the patient of every single cancer cell. For comparison, an antibiotic that is 99.9999% effective would still leave one bacterium in a million alive. Typically, the immune system can deal with this level of bacteria, but this is not necessarily true for cancer cells as even one cell can metastasize to reestablish itself in the body.

Traditionally, there are three forms of treatment for cancer. The following is a discussion of those three forms and also a brief look at some experimental approaches:

1. *Surgery* to remove cancers is effective for centralized tumors and tissues in the immediate vicinity. The difficulty is in removing all of the cancer cells.
2. *Radiotherapy* consists of powerful radiation (X-ray) exposure of the involved area to shrink the tumor. The difficulty is to eradicate 100% of the cancer cells without killing the patient, and the prognosis or outcome of therapy is poor.
3. *Chemotherapy* involves treating the patient by introducing powerful drugs to interfere with reproduction of the fast-multiplying cancer cells.
4. *Experimental approaches* include (a) the development of specific antibodies to selectively destroy cancer cells, (b) laboratory development with later injection of lymphonic-activate killer cells (LAK) to launch all-out war on cancer cells, (c) production of biological response modifiers, such as interferon, to help decode scrambled genetic messages and control the growth of other cells, and (d) antineoplaston therapy.

The real solution is not therapy but taking the preventive actions. People should do everything possible to avoid activities with high risk factors, detect the symptoms early, obtain medical care, and overcome major health threats.

TERATOGENICITY

Teratology is the study of birth defects. Teratogens are chemicals found to have direct links with congenital malformation. The term *teratogens* literally means "monster-making" and refers to the production of abnormal offspring. Research in this field has been dynamic in the last 40 years, and during this time the embryos of mammals, including humans, were found to be susceptible to many (a) physical agents, such as nutritional deficiencies and infections from bacteria or viruses, and (b) chemical agents. It is known that the exact nature of the deformity will vary with the substance that caused it and the time during gestation when it is delivered. To be labeled as a teratogen, it must conform to the following guidelines:

▶ It must significantly increase the occurrence of structural or functional abnormalities.

> Either the mother must be exposed to the substance during the pregnancy or the developing embryo is exposed directly.

> The exposure must be at levels that do not induce severe toxicity in the mother.

The teratogen, by definition then, is different from a mutagen in that there has to be a developing embryo or fetus to be adversely affected. This is a significant and an important issue for pregnant women in the workplace. The following discussion will explain how critical it is to understand the effects of teratogens on pregnancy.

The Developing Fetus

The reproductive system consists of the *gonads,* the testes in the male or the ovaries in the female, and a series of tubes that allow egress of either the sperm from the male or the eggs from the female. A fertilized egg is called a *zygote.* The zygote becomes an *embryo* once it has migrated to the uterus and become nested into its spongy lining. The embryo becomes encased within the *amniotic membrane* and floats freely in amniotic fluid. A *placenta* is formed to deliver nutrients extracted from the mother's bloodstream to the embryo via the umbilical cord. Within the first month, the embryo develops at an amazing speed, becoming over 10,000 times its original size. Different cell layers, tissues, and organs group themselves according to directions from the genes (genetic code). During the second month, the embryo continues to develop under directions from its genes. In the second month, the embryo develops both body segments, an abdomen, a head, and limb buds. Because all the major structures are outlined after 7 or 8 weeks, the stage is entered in which it, in medical terms, becomes a *fetus.* By the end of the third month, the fetus is completely formed with a circulatory system; the "buds" for the limbs are now arms and legs, and the sexual organs and other major organs and systems of the body have appeared. This period of *organogenesis* of the embryo/fetus is known as the *first trimester* (3 months) of the pregnancy.

The embryo/fetus is protected by the placental mechanisms that filter or remove toxic substances from the fetal blood supply. However, the developing embryo/fetus is exquisitely sensitive to foreign or harmful substances because it is developing and growing so rapidly. It absorbs and assimilates large amounts of nutrients from the mother's blood very efficiently. Unfortunately, it can also assimilate many toxins such as caffeine, alcohol, or cocaine. Table 4–2 provides a time line of events that may occur during development of the embryo in the uterus.

The effects of toxic exposure depend on the particular period of embryo/fetal development. Exposure during the first week of pregnancy prevents proper implantation of the embryo in the uterine wall, which causes embryonic death or spontaneous abortion. During the first trimester, an exposure can interfere with limb "bud" formation, such as exposure from thalidomide. Exposure during the third and fourth months leads to neurological, developmental, and behavioral defects. During the fifth and sixth months of pregnancy, exposure leads to endocrine and immune system dysfunction. After seven months of pregnancy, exposure leads to the development of cancer. The kind and variety of abnormality that is produced depends on which organ system is undergoing the most rapid development at the time of the exposure.

It can be extremely difficult to establish cause-and-effect relationships between

▶ TABLE 4–2
Induction of teratogenic effects.

TIME	EFFECT ON DEVELOPMENT
First Trimester	
Week 1	Abortion
Month 1	
Month 2	Organ defects
	Terata
Second Trimester	
Month 4	Behavioral
	Neurological
	Developmental
Month 5	Endocrine
	Immune
Third Trimester	
Month 7	
Month 8	Cancer
Month 9	

a teratogen and the birth defect that it causes. Many years are required before researchers can predict with certainty what substances hold what level of risk to the unborn infant. What is known is that the embryo/fetus is much more susceptible to physical and chemical teratogens than the mother and, therefore, all agents considered to be toxic to the female reproductive system are potential teratogens. Let us review some toxins in this category, as shown in Table 4–3.

Guidelines for Working Women

Birth defects of the embryo/fetus follow the same dose-response relationship as for other populations, but due to increased sensitivity, these are observed at lower doses. However, it is not possible to assign individual risk from such a generalized relationship and, therefore, an expectant mother cannot assume that her unborn baby is safe just because she herself has below-average no-observable-adverse-effect levels

▶ TABLE 4–3
Teratogens and their effects.

Teratogen	Effects
Actinomycin	Can inhibit the formation of RNA
Colchicine	Inhibits cell division
Heavy metals	Inhibits enzyme activity
Diethylstilbestrol (DES)	Can cause cancerous tissue growth in the vagina/scrotum
Alcohol	Causes learning disabilities and psychomotor dysfunction
Alkylating agents	Interrupts DNA synthesis, which causes mutations
Vitamin and mineral deficiencies	Interferes with early organogenesis
Tobacco smoke	Causes same effects as those experienced by the mother
Lead (cosmetics)	Causes defects in central nervous system

Normal woman—normal pregnancy. A normal woman who is experiencing a normal pregnancy should be given work that is neither strenuous nor potentially hazardous (the latter term refers both to the physical hazards and external agents in the work environment that constitute a significant risk). The most common jobs in this class are clerical and administrative, but include many craft, trade, and professional positions. The usual limiting factor on the work that is "strenuous and/or potentially hazardous" to a pregnant woman., fetus, or both is the pregnancy itself as acceptable limits have been established in most positions for the nonpregnant worker. (See discussion on lead exposure later in this chapter.)

Normal woman—abnormal pregnancy. If an abnormal pregnancy is known to exist, a list of the abnormalities of the pregnancy should be developed and comments made on the ability to work in each instance, identifying the specific risk. Included in the evaluation should be the appropriate intervals for the follow-up examinations while working.

Abnormal woman—normal pregnancy. An abnormal woman as used here is one who has a health condition such as heart disease, diabetes, thyroid disease, or multiple sclerosis. Guidelines should be developed for proper evaluation and monitoring during pregnancy while working.

Abnormal woman—abnormal pregnancy. For this condition, the guidelines for the last two conditions should be used.

After childbirth "disability." A normal woman who had a normal delivery should work only in nonstrenous and nonhazardous jobs beginning about 4–6 weeks after delivery. If the worman is abnormal or the delivery is complicated, a list of complications should be developed with an explanation of how each affects the ability to work.

Hazardous exposure. Without question, a pregnant woman should avoid unnecessary exposure to any of the following:
- ionizing radiation
- chemical substances that are mutagenic, teratogenic, or abortifacient
- biological agents of potential harm.

("Unnecessary exposures" may be construed in this context to be regularly at, or above, the "action level.")

▶ **FIGURE 4–5**
Guidelines for assessing hazards in the workplace.
Note. From American College of Obstetricians and Gynecologists and the National Institute for Occupational Safety & Health (NIOSH).

(NOAEL) of exposure. These factors lead to problems in setting occupational health standards where there are women in the workforce, especially when they are pregnant. ("Safe" workplace values have, typically, been designed for the average healthy working adult male.)

Figure 4–5 contains guidelines set forth by the joint efforts of the American College of Obstetricians and Gynecologists and the National Institute for Occupational Safety & Health (NIOSH).

HAZARD RISK MANAGEMENT

Probably the single greatest fear that people have about exposure to hazardous chemicals is cancer. In recent years several significant acts of federal legislation concerning the environment have provided protection from toxic substances, including suspected and known carcinogens.

Government Regulations

The Toxic Substances Control Act (TSCA)

The requirements of TSCA require industries to prevent unreasonable risks to the public by implementing front-end controls and by allowing regulation at any stage in the life-cycle of the drug—from manufacture to disposal—to enforce those controls. The scheme is to evaluate chemicals and institute appropriate controls prior to their commercialization.

The Federal Food Drug & Cosmetics Act (FFDCA)

The original act and later amendments essentially require drug manufacturers to prove the effectiveness of their products to the FDA before they can be marketed.

The Consumer Product Safety Act (CPSA)

When the Consumer Product Safety Commission was developed in 1973 in response to the passage of the Consumer Product Safety Act (CPSA), it established a means for consumers to voice their collective concerns without first having to involve the courts. The act provides standards for regulating flammable products and household chemicals such as paint products, insecticides, and cleaning compounds. The CPSA also provides for consumer voice in such things as involuntary exposure to air or water pollution from sources such as asbestos, radiation, pesticide residues, and formaldehyde.

The Occupational Safety & Health Act (OSHA)

People who have been working for long periods of time in occupations in which they handle chemicals will exhibit health effects from chemicals since they have been exposed to higher concentrations and for longer periods than the general public. One of the primary goals of OSHA is the prevention of all occupational diseases, including cancer. Many modern industrial hygienists and toxicologists postulate that, if a substance has not caused cancer in an exposed occupational group, then it is not likely that it will cause cancer in the general population.

Other Acts of Congress Enforced by the Environmental Protection Agency (EPA) and Other Federal or State Agencies

Toxicologists and other health professionals recognize that certain "environmental" factors are present as various background levels of contamination. The EPA and state-run environmental programs are enabled by a variety of legislation and regulations. These include (a) The Clean Air Act, (b) The Clean Water Act, (c) The Safe Drinking Water Act, (d) The Toxic Substance Control Act (TSCA is an EPA Act), (e) different radiation controls in conjunction with the Nuclear Regulatory Commission and the Department of Energy, and (f) Resource Conservation and Recovery Act (RCRA) in 1976, which covers management of hazardous waste from cradle to grave. Hazardous materials are also governed while they are being transported by the U.S. Department of Transportation.

An Interdisciplinary Approach

In the future, reducing the risk of exposure to physical and chemical carcinogens is going to require an interdisciplinary approach. This means that the various government agencies, private health organizations, and the scientists need to combine their efforts. Figure 4–6 is a compilation of the key players in these efforts!

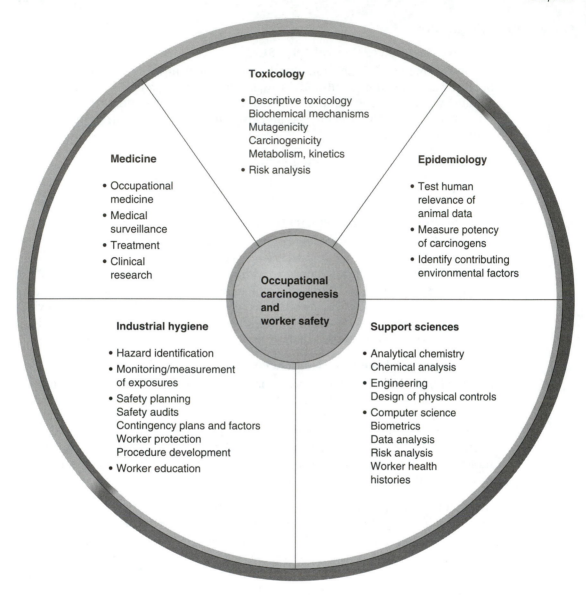

▶ **FIGURE 4–6**
Identifying and reducing chemical carcinogens requires an interdisciplinary approach in which health professions work closely with other scientific disciplines.

SUMMARY

▶ This chapter discussed the functions of the cell and the role that genes play in controlling cellular activities and reproduction.

▶ Chemicals pose a potential risk to human reproduction in a variety of ways: by mutagenesis, carcinogenesis, and teratogenesis.

▶ Within the cell are small organelles that perform certain functions. These are managed by messages encoded with the nucleic acids of RNA and DNA, and are called the genetic code. The messages cause to be created long strings of amino acids, each with its own sequence, known as a polypeptide.

▶ Very long strings of such polypeptides are called proteins. A functioning protein is called an enzyme. Enzymes are responsible for performing all of the cell's upkeep and maintenance.

▶ When a mutation occurs, it results in an alteration of the message. The chemical or agent that causes this mutation is known as a mutagen.

▶ While some mutagens are found everywhere in our environment and may even be useful as an evolutionary mechanism, many cause much suffering and misery because they can lead to cancer and other maladies.

▶ The body (somatic) cell has natural repair mechanisms that can recognize and intercept a faulty genetic code to prevent the effects that would have resulted from this faulty message.

▶ This chapter provided a discussion of (a) carcinogenesis and the primary factors of lifestyle, environmental background, and exposure; (b) its common characteristics of cell proliferation, loss of differentiation, and metastasis; (c) the different periods involving initiation and latency and the types of carcinogens recognized as initiators (genotoxic) and as promoters (epigenetic); (d) the main four types of cancers, which include leukemia, lymphomas, sarcomas, and carcinomas; (e) cancers associated with essential target organs such as lung, liver, urinary tract, skin, breast, colon, and oral cancers; and, finally, (f) the methods of therapy for cancer such as surgery, radiotherapy, chemotherapy, and several experimental approaches.

▶ The developing embryo is particularly susceptible to both physical and chemical hazards. When the effects are adverse, the process is called teratogenicity.

▶ There are different stages of reproduction involving conception and development of the embryo/fetus.

▶ The American College of Obstetricians and Gynecologists and the National Institute for Occupational Safety and Health (NIOSH) have issued guidelines for working mothers.

▶ Many people in the United States have a phobia about cancer and are concerned about government efforts that regulate the presence of cancer-causing substances and agents (carcinogens) in our environment.

STUDY QUESTIONS

1. What is the cellular role of the nucleic acids? Identify the two principal forms.
2. What is meant by the term *genetic code?*
3. What is a mutagen? Provide a brief summary of their desirable and undesirable effects.
4. What are the primary causes of carcinomas or tumors? Explain why some are malignant and others are benign.
5. What is the meaning of *carcinogenic?*
6. What are the different types of cancer and the toxic substances that cause or are suspected of causing them?
7. What is the meaning of *teratogen?*
8. At what stage in reproduction is the embryo/ fetus most affected and by which toxic substances? Describe the stages of growth of the unborn baby.
9. What regulatory acts and efforts does the government make to protect general citizens and occupational workers from the risk of exposure to toxic substances?
10. How should future efforts be coordinated for cancer research and treatment?

5

Systemic Toxicology-1

COMPETENCY STATEMENTS

This chapter provides a brief introduction to the principles of systemic toxicology. After reading and studying this chapter, students should be able to meet the following objectives:

▶ Define the meaning of systemic toxicology and its relationship to the primary target organs of the body.

▶ Demonstrate a basic understanding of the following conditions and substances: hypoxia, anemia, cardiotoxins, hematoxins, neurotoxins, and reprotoxins.

▶ Describe the purpose and the function of the integumentary system and its relationship to systemic toxicology.

▶ Describe the purpose and the function of the blood and its relationship to systemic toxicology.

▶ Describe the purpose and the function of the cardiovascular system and its relationship to systemic toxicology.

▶ Describe the purpose and the function of the endocrine system and its relationship to systemic toxicology.

▶ Describe the purpose and the function of the central nervous system and its relationship to systemic toxicology.

▶ Describe the purpose and the function of the reproductive system and its relationship to systemic toxicology.

INTRODUCTION

Paracelsus observed that the dose determines if a substance acts as a poison. Therefore it is misleading to label a substance as "toxic" or "nontoxic" without specifying the details relating to the dose (the what, where, when, and how of the exposure as discussed in Chapter 1). Many substances in the environment are necessary to human function and are essential ingredients to survival and life itself. For example, there are numerous metals that are found in the body and need to be replaced in minute amounts, such as zinc, magnesium, arsenic, selenium, and copper. These same elements may be toxic, and even deadly, at higher doses than are needed by the body for normal daily operations. However, until these substances actually enter the body and reach a "target" organ, where they cause their damage, these substances display no toxic characteristic. Toxicity, then, is not intrinsic to the substance itself but, rather, depends on the effect it has on the body.

In a previous chapter, we discussed the various routes by which chemicals enter the body. In this chapter we will discuss *systemic toxicology,* or the study of what happens when these chemicals find their way to various tissues, organs, and organ systems of the body, in other words, their target organs.

Physiology is the study of how the body and its parts work or function, whereas *anatomy* is the study of the structure and shape of the body and body parts and their relationship to one another. Each of these areas is a subdivision of biology which, more simply put, describes how our bodies are put together and how they work. Figure 5–1 shows the hierarchy of living matter. The simplest level of structure is *atoms,* which are tiny building blocks that combine to form *molecules.* These may be simple, such as water, or complex, such as sugar, fats, nucleic acids, and proteins. These complex biological molecules associate in certain ways to form a microscopic *cell,* which is the smallest unit of all living things. Cells vary widely in size, shape, and function.

Similar cells associate to form *tissues,* which are groups of cells with a particular function. These include the following:

1. *Epithelial tissues:* These tissues line or cover the body and have functions such as protection, filtration, absorption, excretion, and secretion.
2. *Connective tissues:* These tissues are those that hold the body together, such as bone, cartilage, and fat.
3. *Muscle tissues:* These tissues move the bones, operate the heart, and line the walls of hollow organs such as the stomach, bladder, and blood vessels.
4. *Nervous tissues:* These tissues are an association of cells we call *neurons,* which conduct information via electrochemical impulses from one part of the body to another.
5. *Blood:* Although, technically, blood is a connective tissue, it is unique among all of the body's tissues because of its fluid nature and therefore deserves special consideration.

An *organ* is a structure that is composed of two or more tissue types and has a specific body function. All of the body's organs of similar purpose are grouped into an *organ system* that allows them to cooperate to accomplish a common purpose. The digestive system, which connects the mouth to the anus, links the esophagus (throat) to the stomach, to the duodenum, and then to the rest of the

▶ **FIGURE 5–1**
Hierarchy of living matter.

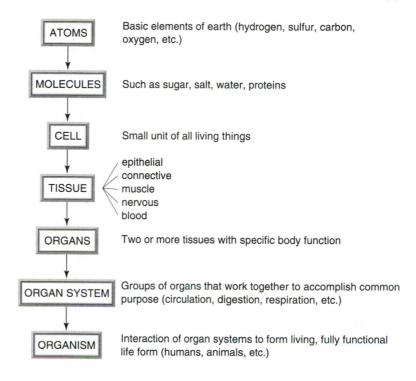

ATOMS — Basic elements of earth (hydrogen, sulfur, carbon, oxygen, etc.)

MOLECULES — Such as sugar, salt, water, proteins

CELL — Small unit of all living things

TISSUE — epithelial / connective / muscle / nervous / blood

ORGANS — Two or more tissues with specific body function

ORGAN SYSTEM — Groups of organs that work together to accomplish common purpose (circulation, digestion, respiration, etc.)

ORGANISM — Interaction of organ systems to form living, fully functional life form (humans, animals, etc.)

small and the large intestines. It is an example of an organ system and is known as the gastrointestinal (GI) tract, the digestive system, or the gut. The digestive system is basically a tube that runs through the body from the mouth to the anus, which includes the oral cavity at one end of the system and the rectum and anus at the other.

Depending on how you sort "systems," we, as an *organism,* are comprised of 11 organ "systems" that interact and allow us to exist as living, breathing, fully functioning human beings or *human organisms.* In this chapter we will discuss the following organ systems and highlight how they are the targets of hazardous substances:

1. Integumentary system (skin)
2. Blood
3. Cardiovascular system
4. Endocrine system
5. Central nervous system (CNS)
6. Reproductive system

In Chapter 6 the subject of systemic toxicology continues with the study of the following organs and organ systems.

1. Respiratory system
2. Liver
3. Kidneys
4. Eyes
5. Immune system

THE SKIN

The *cutaneous membrane,* superficially made up entirely of epithelial cells on a deeper layer of connective tissue, is generally called the *skin.* Technically, the *integumentary system,* in reality, also includes, in addition to the "skin," a number of organs and glands such as the *sebaceous (oil) glands,* the *sudoriferous (sweat) glands,* hairs, and nails. This system is much more than just a covering for the body; structurally, it is a marvel! The skin interfaces with the external environment and (a) keeps water and other precious nutrients *in* the body; it also (b) keeps water and other unwanted things *out* of the body. The skin thus constitutes the first line of defense as a barrier to chemicals that are exogenous to the body's functions.

Basic Structure

The *epidermis* is the outer layer of the skin, which has no blood supply of its own. On its uppermost layer is the *stratum corneum.* This "cornified" layer is hardened to prevent water loss from the body surface and consists of 20 to 30 shingle-like layers of flat, lifeless cells. The body sheds several layers of the cells every day, and by the time these cells reach the free surface, all that is left is a flake of *keratin,* a horny material that constitutes an excellent barrier to many chemical agents and microorganisms. However, many solvents easily dissolve and penetrate this protective barrier. The underlying *dermis* is made up of dense fibrous and elastic connective tissue, which is the strong, stretchy envelope or "hide" that helps hold the body together. The two layers are cemented firmly together to form the tough structure of the skin. The deepest reticulated skin contains blood vessels, hair, and sweat and oil glands and is rich in nerve supply, pain, touch, hot, cold, and deep pressure receptors. Under the skin lies the *subcutaneous tissue,* which is not regarded as part of the skin but anchors the skin to underlying muscle, bone, and other organs.

Although the skin is the body's first line of defense, there are many less well-protected entry points along the skin such as the glands and hairs mentioned earlier. These entry points are in a reticulated layer, but they have avenues to the exterior of the body, such as hair from follicles, which extend beyond the outermost surface of the skin. As can be seen in Figure 2.9, there are numerous routes by which a toxic substance can be absorbed through or penetrate the skin. Generally, the skin serves as an efficient barrier, and any substance must pass through a thick layer of epithelial cells, as opposed to most other organs of the body, which may have a barrier of only one or two epithelial cells in thickness. (Figure 2–9 illustrates the components of the integumentary system.)

Functions of the Skin

Most of the skin's functions are protective, in that it:

1. Insulates and cushions the deeper organs of the body
2. Protects the body from:
   a. Mechanical damage such as bumps and grinds
   b. Chemical damage, such as burns from acids and bases
   c. Thermal damage, such as from heat or from cold
3. Regulates body temperature
4. Forms a barrier to, and habitat for, exogenous substances and materials like bacteria

Those of us in good health possess a natural coat that is waterproof; permanent press; repairs its own cuts, burns, and rips; is stretchable and washable; and, if we take care of it, will last a lifetime.

Physiology of Absorption

Lipophilic substances like paint thinner and organics pass across the epidermis layer by means of *passive diffusion.* This means the process occurs without need of any additional energy, other than their own energy of motion or *kinetic energy.* In humans there are differences in permeability from body area to body area as the thickness and composition of the skin varies. For example, the skin on the soles of the feet and palms of the hands is much thicker than the skin on the eye lids, the inner arms, and the thighs.

The second phase of penetration is through the dermis layer. This occurs by diffusion into the watery interstitial fluid medium and into the circulatory system. The rate of such dermal absorption is influenced by a large number of environmental, behavioral, and physiological factors. Environmental factors include heat, relative humidity, wind vibration, ultraviolet and visible radiation, and electrical current. Other factors that affect absorption rates are safety precautions, individual health status, and genetic factors.

The skin itself contains many protective mechanisms. These include sweating, secretion of oils, phagocytic cells, metabolic detoxification, and pigmentation (to protect from radiation). The skin is capable of some biotransformation reactions. A person's hygiene, habits, age, gender, and medications all influence susceptibility and degree of response to insult of the skin. The properties of the substance itself must also be considered. Lipophilic substances, such as organophosphate insecticides, penetrate the skin quite readily, whereas hydrophilic substances are very slow because of the skin's resistance to water and need to find the weak spots in the barrier.

Skin Reactions

The skin is capable of responding to many hundreds of ailments. The most common disorders result from allergies, infections, and burns. Ailments from thermal and chemical burns, irritations from the insult of chemicals, corrosions, photoallergies, and chloracne create a variety of cutaneous reaction patterns.

Infections and Allergies

Many skin disorders result from bacteria, viruses, or fungi. Athlete's foot is a condition in which itchy, scaly skin appears between the toes as a result of a fungal growth. Boils and carbuncles are inflammations of hair follicles, sebaceous glands, and reticulated layers due to bacterial infection. Cold sores are small fluid-filled blisters caused by the herpes simplex virus, and impetigo is a condition of pink, water-filled lesions around the mouth and nose that develop a yellow crust and are caused by an infection from a staphylococcus bacteria. Psoriasis is a chronic condition characterized by reddened lesions covered by dry, silvery scales.

Burns

When the skin is severely damaged, nearly every system of the body reacts. A burn is tissue damage and cell death caused by heat, electricity, UV radiation (sunburn), and certain chemicals. Burns can lead to dehydration, kidney shutdown, and circulatory shock from the loss of body fluids. Acids and bases are examples of chemicals that can cause burns to the skin. Deep chemical burns are discussed under the topic of corrosions.

Irritations

Direct exposure of the skin often produces local cutaneous inflammatory response of the skin, such as *contact dermatitis,* a term that identifies the irritant responses of itching, swelling, and redness of the skin, which leads to blisters resulting from exposure to chemicals such as those in poison ivy or oak. These reactions of the skin can be divided into acute and chronic or cumulative irritation.

Acute Irritation   This injury is the result of a single incident of skin contamination and is produced by a relatively large number of substances. Examples include strong solvents and corrosives such as acids and bases. Damages that result may or may not be reversible depending on the extent and depth of tissue damage.

Chronic or Cumulative Irritation   This type of injury results from continuous and repeated exposures. While these exposures alone do not produce an acute response, eventually they "accumulate" and cause inflammation and also promote susceptibility of the skin to other chemicals. A substance that eventually leads to a chronic reaction is known as a *marginal irritant.* Examples of such irritants are:

1. *Corrosions:* Direct chemical action on the skin, or on any tissue, may produce ulcers, disintegration, necrosis, and irreversible damage. These reactions are caused by extended contact when the person exposed fails to immediately flush the area with water for at least 15 minutes following exposure. It is similar to a deep third-degree burn and occurs from contact with extremely corrosive acids and bases, such as hydrochloric and sulfuric acid and sodium hydroxide or potassium hydroxide. They also occur from exposure to strong solvents, such as carbon tetrachloride, toluene, and a wide variety of alkalis and phenolics, which can rapidly eat away the skin.

2. *Photoallergies, sensitivity, and toxicity:* Photoallergies are induced by photochemical alterations of substances to products that elicit a specific immune response. Subsequent exposure to these substances, in the sunlight, induces allergic reactions. Substances that are capable of producing these effects include the drugs griseofulvin and promethazine; deodorants and disinfectants such as tetrachlorsalicylanilide, hexachlorophene, and bithionol; sunblocks, aminobenzoic acid and digaloyltrioleate; and blancophores and wash whiteners.

    Photosensitivity is irritation caused by light in conjunction with a chemical exposure. Washing whiteners and furocoumarins, psoralenes, are photosensitizers. After exposure there is an increased pigmentation of the skin, typical of a suntan; however, the tan is frequently blotchy and discolored. Unfortunately, the sensitization can remain long after exposure to the chemical has ceased.

    Phototoxic reactions occur when a substance absorbed locally by the skin is biotransformed into a toxin by a photochemical reaction. The nature of the response depends on the nature of the toxin.

3. *Chloracne:* Chloracne lesions, or blackheads, formed on the skin, frequently progress to be pustules and abscesses. Agents that produce such inflammation

are greases and oils, coal tar pitch, creosote, ingredients in many cosmetics, and poisoning by aromatic hydrocarbons. The highest incidence of chloracne results from exposure to TCDD (tetrachlorodibenzo-p-dioxin), which can bring on nausea, headache, vomiting, and even changes in the excretory system; but first there is a tendency to exhibit profuse acne. Other substances implicated in the production of chloracne are polyhalogenated naphthalene, biphenyls (PCBs), dibenzofurans, and dioxins.

4. *Cutaneous reaction patterns:* A number of characteristic cutaneous reaction patterns, besides those discussed earlier, are known. The following are examples:

   ▶ Physical dermatitis, as from fiberglass
   ▶ Urticarial reactions, wheals and hives, resulting from biogenic polymers
   ▶ Cutaneous granulomas, often resulting from beryllium, zirconium, chromium salts, and silica
   ▶ Hair damage and loss from alkali, thioglycolates, and oxidizing agents such as peroxides
   ▶ Hypopigmentation, loss of color from phenols and catechol including hydroquinone, butyl phenol, and menobenzylether
   ▶ Hyperpigmentation, increase in color due to phenolphthalein; barbiturates; heavy metals such as bismuth, arsenic, mercury; antimalarials; antibiotics such as tetracycline; and alkylating agents
   ▶ Cancer of the skin may be caused by ionizing radiation, polycyclic aromatic hydrocarbons, arsenic, and a number of other agents that affect the carcinogenicity of UV radiation

The effects of the exposure of our skin to the sun tend to be negative. The medical and public health communities have made concerted efforts to discourage us from excessively exposing our skin to natural sunlight. To this end, it is recommended that dry clothing be worn (wet cloth is marginally effective in blocking UV radiation); a hat should be worn to shade the face and ears. Sunblock should be applied to exposed skin surfaces. Glass or UV-blocking sunglasses should be worn.

Conclusion

The skin is one of the body's major organ systems, representing about 10% of body weight. The skin is important to the function, metabolism, and integrity of the whole body. A wide variety of toxic agents may contact the skin in both the workplace and in our everyday activities. Because of this potential exposure, it is clearly important to understand the dangers and hazards associated with these substances. Being aware is essential in maintaining health and safety.

THE BLOOD

Blood is a complex connective tissue that serves as the medium for transporting oxygen, nutrients, hormones, and other substances. The transport is made to and from the body's cells where exchanges are made and from which wastes are removed. Blood is the "river of life" that surges within us and transports almost everything that must be carried from one place to another within the body to sustain life.

Functions of the Blood

Table 5–1 summarizes the functions of the blood, which include the following:

1. Supplies oxygen and nutrients to the cells of the body
2. Transports wastes to excretory organs for elimination
3. Defends against damage from injury and disease
4. Regulates body temperature by distributing body heat
5. Transports hormones to cells
6. Maintains the acid/base balance, or pH, of the tissues
7. Forms clots to maintain fluids within the body

Composition of the Blood

Blood accounts for approximately 8% of body weight, and in a healthy adult the total volume is approximately 5 to 6 liters (6 quarts). The blood has two components: the liquid part of the blood called *plasma* and the living blood cells called the *formed elements*. Figure 5–2 illustrates these two components of blood.

Plasma

This fluid portion of the blood is approximately 90% water, and many different substances are dissolved in this fluid. Hundreds of organic and inorganic substances are either dissolved or suspended in the blood. These include electrolytes, various nutrients, blood sugar (glucose), and a large number of proteins such as albumins, globulins, and fibrinogen, which are essential to transportation and, in the case of fibrinogen, to the formation of blood clots. There are also respiratory gases, hormones, cellular wastes, by-products of cellular metabolism, and many other exogenous materials. The composition varies continuously as cells remove or add substances to the blood.

Fluid circulates within the cardiovascular system as plasma, in between the cells as interstitial fluid, and within the lymphatic system as lymph. A portion of the liquid fraction of blood filters out of the circulatory system into the spaces surrounding the cells, called interstitial space. This filtered blood is called *interstitial fluid,* and it contains both nutrients and waste products of the cell and any other substance transported by the circulatory system. The interstitial fluid enters into one end of the lymphatic system through tiny structures called *blind capillaries,* and this fluid portion is now, technically, called lymph. Once the lymph has been processed, it is disgorged at the other end of the bloodstream. Thus, fluid flows from one reservoir to another. We will discuss this more in-depth when we describe the immune system.

Formed Elements

Suspended in the plasma, or nonliving fluid, are living blood cells and corpuscles called the formed elements. They make up between 45% and 55% of the whole blood. The three types of formed elements are erythrocytes, leukocytes, and platelets.

Erythrocytes *Erythrocytes* function primarily to transport oxygen in blood to all cells of the body. These formed elements are also known as the *red blood corpuscles (RBCs),* which are sacs of h*emoglobin (Hb)* molecules that form bonds between their iron atoms and atoms of oxygen, which are then carried along in the bloodstream.

▶ TABLE 5–1
A review of formed elements of the blood.

Cell	Abundance (per μl)[a]	Characteristics	Functions	Remarks
RED BLOOD CELLS	5.2 million (range 4.4–6.0 million)	Flattened, circular, no nucleus, mitochondria, or ribosomes; red color due to presence of hemoglobin molecules.	Transport oxygen from lungs to tissues, and carbon dioxide from tissues to lungs	120-day life expectancy; amino acids and iron recycled; produced in bone marrow
WHITE BLOOD CELLS	7000 (range: 6000–9000)			
Granulocytes Neutrophils	4150 (range: 1800–7300) differential count: 57%	Round, nucleus resembles a series of beads; cytoplasm contains large, pale inclusions	Phagocytic; engulf pathogens or debris in tissues	Survive minutes to days, depending on activity; produced in bone marrow
Eosinophils	165 (range: 0–700) differential count: 2.4%	Round, nucleus usually in two lobes; cytoplasm contains large granules that stain bright red with acid dyes	Phagocytic: engulf anything in tissues that is labeled with antibodies	Produced in bone marrow
Basophils	44 (range: 0–150) differential count: 0.6%	Round, nucleus usually cannot be seen due to dense, blue granules in cytoplasm	Enter damaged tissues and release histamine and other chemicals	Assist mast cells of tissues in producting inflammation; produced in bone marrow
Agranulocytes Monocytes	456 (range: 200–950) differential count: 6.5%	Very large, kidney bean-shaped nucleus, abundant pale cytoplasm	Enter tissues to become free macrophages; engulf pathogens or debris	Primarily produced in bone marrow
Lymphocytes	2185 (range: 1500–4000) differential count: 31%	Slightly larger than RBC, round nucleus, very little cytoplasm	Cells of lymphatic system, providing defense against specific pathogens or toxins	T cells attack directly; B cells form plasma cells that attack with antibodies; produced in bone marrow and lymphatic tissues
Platelets	350,000 (range: 150,000–500,000)	Cytoplasmic fragments, contain enzymes and proenzymes; no nucleus	Hemostasis: clump together and stick to vessel wall (platelet phase); activate intrinsic pathway of coagulation phase	Produced by megakaryocytes in bone marrow

[a]Differential count: percentage of circulating white blood cells.

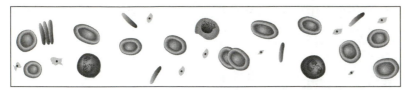

Whole blood

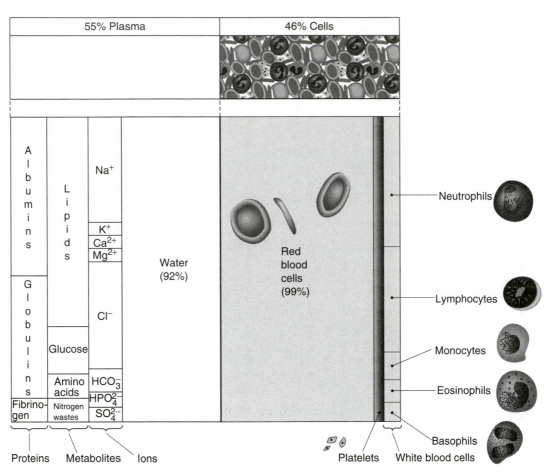

▶ FIGURE 5–2
Composition of whole blood.

RBCs outnumber white blood cells by 1000 to 1 and are the major factor contributing to blood viscosity. A decrease in the oxygen-carrying ability of the blood is known as *anemia* and can result in a lowered count of RBCs or a deficient hemoglobin content in the RBC. *Polycythemia* is an abnormal increase in RBCs and can result from bone marrow transport.

Each single RBC contains about 250 million hemoglobin molecules, and each Hb molecule can bind with 4 molecules of oxygen: the result is that each tiny RBC carries 1 billion molecules of oxygen. RBCs are confined to the bloodstream and carry out their function within the blood itself.

Leukocytes *Leukocytes,* otherwise known as *white blood cells (WBCs),* help defend the body against invasion by exogenous substances. They are thus crucial to the body's defense against disease. The role and function of WBCs is discussed further in the immune system section. WBC cells are produced by either bone tissue or by the tissues of the lymphatic system. In contrast to RBCs, white blood cells are able to slip into and out of the blood vessels into various tissues in a process called *diapedesis.* WBCs use the circulatory system for transportation to various areas of the body where their services are needed.

WBCs consist of two subgroups: The first are the *granulocytes,* which break down further into three types called *neutrophils, eosinophils,* and *basophils.* These cells are named for their ability to be stained when subjected to different Wright's stains such as neutral, acid, and base stains. The second group are the *agranulocytes,* which break down further into two types called *lymphocytes* and *monocytes.* The role and function of these specific types of leukocytes will be discussed further in the section on the immune system.

Platelets *Platelets,* also known as *thrombocytes,* are not cells in the strict sense but are fragments or buds of much larger cells with a multiple nucleus. When these large cells rupture, they scatter 50 or more "pieces" that quickly seal themselves off from the surrounding fluid and appear as irregularly shaped bodies among the other blood cells. Platelets are needed for the clotting process that occurs when blood vessels are ruptured or broken and need repair.

Formation of Red Blood Corpuscles

Blood cell formation, or *hemopoiesis,* occurs chiefly within the red marrow of the flat bones of the body such as the skull, sternum, ribs, and pelvis. Each type is produced in different numbers in response to changing needs of the body. Figure 5–3 illustrates the number and variety of blood cells that are all formed from the common type of stem cell (or mother cell) known as the *hemocytoblast* that resides in the red bone marrow. Before the RBC exits the bone marrow, it loses its nucleus to become a corpuscle. Because it no longer contains a nucleus, it is no longer self-repairing. Therefore, it has a limited "life time," normally 120 days, before it is scavenged and recycled.

Hematoxins

Chemicals that decrease the ability of hemoglobin to deliver oxygen to the body tissues are called *hematoxins.*

Hypoxia

Hypoxia refers to any condition in which there is inadequate oxygen to support the tissue function. A common example of hypoxia is carbon monoxide (CO) poisoning. Sufficient oxygen is unable to find binding sites on hemoglobin molecules because of the high affinity of carbon monoxide to bind to the iron in hemoglobin. Carbon monoxide binds with hemoglobin at a rate approximately 200 times the rate of oxygen and, literally, denies oxygen a place to bind with the blood. The result is that the RBCs are unable to transport sufficient amounts of oxygen to the cells, and oxygen deprivation, or hypoxia, occurs.

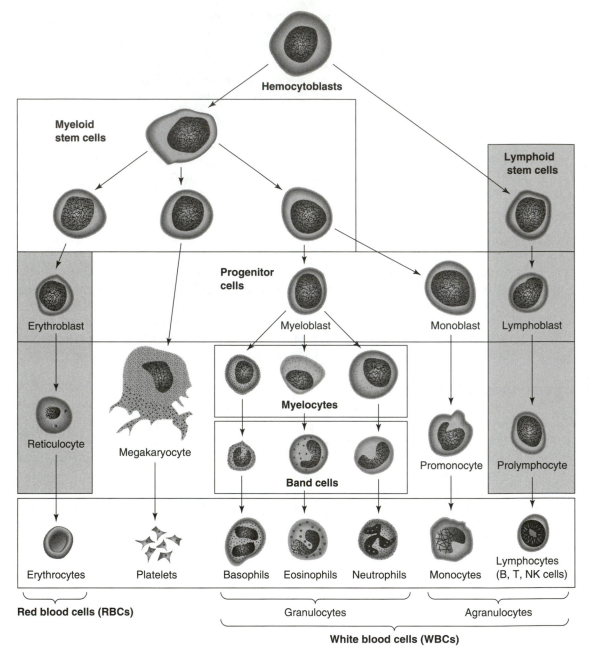

▶ **FIGURE 5–3**
Origins and differentiation of blood cells.

A second example of hypoxia occurs when nitrates, aromatic amines, nitrogen compounds, and chlorate salts bind with iron in the blood. The iron atoms become oxidized and are converted from a ferrous state to a ferric state, which results in the formation of *methemoglobin*. This process both denies hemoglobin the ability to bind oxygen and also interferes with the blood's additional function of removing carbon monoxide. *Sulfhemoglobin* exhibits a similarity to methemoglobin with the exception that this condition results from the presence of sulfur-containing compounds such as the sulfonamides and occurs when the blood is exposed to hydrogen sulfide. Sulfhemoglobin molecules have no place for oxygen to bind and thus have no oxygen-carrying capacity whatsoever.

Histotoxic Hypoxia

Histotoxic Hypoxia occurs when there is an adequate flow of blood and thus an adequate supply of oxygen, but the metabolic processors of the cell (the *mitochondria)* are rendered incapable of using the available oxygen needed for their particular metabolic function. Cyanides act on the mitochondria to prevent the formation of *ATP,* which is the essential source of energy for all cellular activity. A number of substances also uncouple certain chemical bonds and result in exothermic reactions (those that produce heat in the body). Some examples include dinitrophenol, used in the past to treat obesity and resulting in symptoms of high body fever. Several herbicides, such as paraquat and diquat, interfere with the transformation in the cells of hydrogen to a substance called nicotine adenine dinucleotide phosphate *(NADP).* Without NADP the process known as glycolysis, the cellular process of breaking down blood sugar (glucose) into simpler compounds that are useable by the body, simply cannot occur. Interruption of the citric acid cycle, called the *Krebs cycle,* can result in the inability of the cell to liberate, or release, energy from carbohydrates, fats, and amino acids during metabolism. Without energy from the food we eat, the body cannot function.

White Blood Cell Disorders

White blood cell disorders can be either too many WBCs, known as *leukemia,* or from too few, a condition called *leukopenia.* Leukemia is a cancer of the blood, and thrombocytopenia (too few thrombocytes, or platelets) leaves the victim susceptible to uncontrollable hemorrhage. These can be induced by exposure to benzene, bismuth, chlordane, lindane, mercuries, potassium iodine, and toluene.

Other Disorders

Other disorders are associated with a reduction in white blood cells, which leaves the afflicted person susceptible to most infections and a myriad of resulting symptoms. A number of chemicals such as benzene, DDT, dinitrophenol, quinidine, quinine, and the silicylates have been linked to this syndrome. *Anemia,* as mentioned earlier, occurs when the red bone marrow fails to produce an adequate number of red blood cells and may be caused by exposure to alkylating agents, arsenicals, benzene, carbon tetrachloride, chlordane, insecticides, potassium perchlorate, and salicylate. *Hemolytic anemia* can be induced from exposure to such chemicals as arsine, lead, methyl, naphthalene, nitrobenzene, and phenylhydrazine.

Symptoms

Headache, weakness, nausea, dizziness, and flushed skin are all symptoms of blood disorders. Routine blood analysis should be done whenever an individual is accidentally, potentially, or chronically exposed to chemicals.

THE CARDIOVASCULAR SYSTEM

The *circulatory system* actually consists of two separate and distinct organ systems: the cardiovascular system and the *lymphatic system*. The primary organs of the *cardiovascular system,* which will be discussed in this section, include the heart and the blood vessels.

Figure 5–4 illustrates the heart and its primary delivery tubes to and from various parts of the body. *Arteries,* the large tubes that carry blood from the heart, branch into smaller and smaller tubes and eventually become *arterioles.* These small tubes have special smooth muscular restrictions and valves that ultimately regulate blood flow. The arterioles feed *capillary beds* that intimately intertwine with the cells and have a transparent wall that is just one cell in thickness. Additionally, small spaces between cells, called fenestration, allow for leakage of fluid. These structures allow for an easy exchange between these very small tubes that permit the plasma and gases to permeate

❱ **FIGURE 5–4**
The cardiovascular system:
The heart.

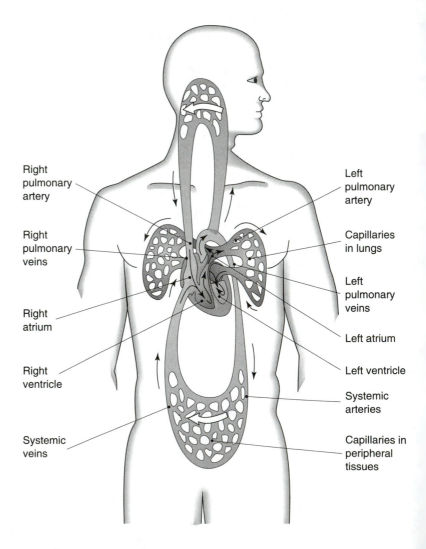

into the interstitial spaces of the cells. This interstitial fluid is then redirected out of the spaces and is collected by another set of little tubes, the *venules,* which direct the blood plasma back toward the heart. These smaller tubes coalesce into larger tubes called *veins* and they, in turn, empty the blood back into the heart on its return trip. This system of tubes throughout the body is called the *vascular structure.*

Two types of controlling systems regulate heart activity. One of these is the restriction of smooth muscles in the walls of the arteries, which act like "brakes" and "accelerators" to control the rate of blood flow. These *cardiac muscle cells* expand and contract spontaneously and independently even if all nervous systems are disconnected. The second control system is the *intrinsic,* or *nodal, system* that is built directly into the tissue of the heart itself. The *SA node,* or *sinatorial node,* is often called the "pacemaker" of the heart and literally directs the rate at which the heart beats. Impulses are sent from the SA node through specialized cells of the atria to the *AV node,* or *atrioventricular node,* and then the atria contracts. The impulse is delayed briefly at the AV node to give the atria time to finish contracting. The impulse is then passed quickly through the AV bundles and branches through the muscles in the ventricles, through cells called *Purkinje fibers,* which results in a "wringing" contraction (similar to milking a cow) that, when functioning properly, efficiently forces the blood out of the ventricles of the heart and into the arteries (Figure 5–5).

Chemicals can selectively target the heart muscle tissue, which is called the *myocardium.* In addition to the direct adverse effects of toxic chemicals on the myocardium, there is also a danger that the conducting systems of the heart will be disrupted. Frequently, the lethal risk of failure is greater after an exposure because of the possibility of an irregular heartbeat, called *arrhythmia,* which makes it difficult for the heart to work efficiently. If the impulses regulating the heart are not regular, then the contractions are out of rhythm, and since these impulses are chemical in nature, they may be structurally altered after even a single exposure to specific substances. Arrhythmia can be the result of a number of different conditions: (a) alterations of the impulse rate, (b) disturbance of the site of impulse origin, (c) loss of force or strength and shape of contraction of the impulse itself, and (d) other disorders of the circulatory function. Each of these conditions and its causes is discussed below.

The heart is also susceptible to secondary effects of damage to other organs of the body that regulate and stabilize the physiological state of the body called *homeostasis.* Changes in the *pH balance,* those changes in the balance and composition of electrolytes, and changes in the balance of oxygen can each have a negative impact on the heart. Loss of oxygen to an area of the heart muscle may cause cell death and result in a *myocardial infarction,* more commonly called a heart attack.

Alterations of Impulse Rate

Impulses originate in the heart by depolarization (discussed further in the section on the central nervous system) of the Purkinje fibers located in the SA node. The rate of depolarization is somewhat under the automatic, or involuntary, control of the autonomic nervous system (ANS). The relative stability of our internal environment depends largely on the workings of the ANS, and certainly the heart beat rate can be altered by changes to the chemical balance within the ANS. It is important for the heart to return to a normal "resting state" and get ready, in time, for another impulse. The impulse itself is carried through a "slow channel," by strontium and barium ions

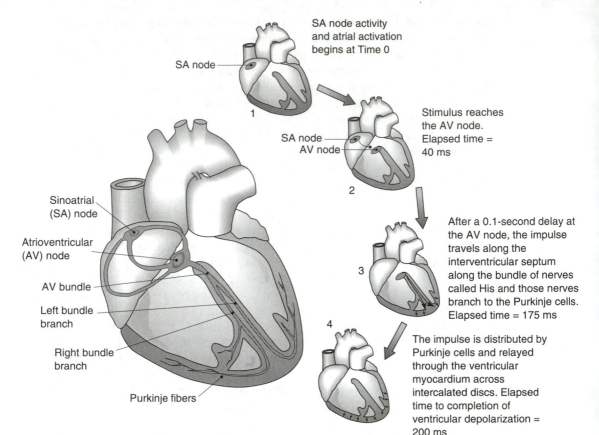

SA node activity
and atrial activation
begins at Time 0

SA node

1

SA node
AV node

Stimulus reaches
the AV node.
Elapsed time =
40 ms

2

Sinoatrial
(SA) node

Atrioventricular
(AV) node

AV bundle

Left bundle
branch

Right bundle
branch

Purkinje fibers

3

After a 0.1-second delay at
the AV node, the impulse
travels along the
interventricular septum
along the bundle of nerves
called His and those nerves
branch to the Purkinje cells.
Elapsed time = 175 ms

4

The impulse is distributed by
Purkinje cells and relayed
through the ventricular
myocardium across
intercalated discs. Elapsed
time to completion of
ventricular depolarization =
200 ms

▶ FIGURE 5–5
The conducting system of the heart.

rather than through normal calcium channels, to allow the heart this needed time to prepare. Effects on the initiation or origination of the impulse may therefore be caused by altering these chemical balances.

Heart rate can also be affected by many different conditions. *Tachycardia* is a rapid heart rate (over 100 beats per minute) while at rest, and *bradycardia* is a heart rate that is substantially lower than normal (less than 60 beats per minute) while at rest. Although, by themselves, neither condition is pathological, prolonged tachycardia can progress to *fibrillation,* a rapid uncoordinated shuddering of the heart muscle that makes the heart totally useless as a pump. Fibrillation is the major cause of death from heart attacks in adults.

Disturbance of the Site of Impulse Origin

The AV node, in general, is the most common site of conduction disturbances in the heart. These disturbances are usually due to continuous impulses for one or more cycles of the heart. If the pace of the heart beat becomes too rapid, the ability of the myocardial cells to respond in a coordinated manner becomes lost and the muscles start to "flutter" or become spastic. The result is that the coordinated contraction necessary for a good wringing of the heart and pumping of the blood is also lost. Once this overall coordination is lost, individual muscle cells may start to depolarize on their own in a condition known as *ectopic foci.* Instead of the AV acting as a single pacemaker directing the beat of the heart, the depolarizing muscle cells become a number of uncoordinated pacemakers.

Loss of Force of Contraction

Contraction of the cardiac muscle is initiated first by depolarization of the cell membrane, and subsequently, the cell is repolarized to contract to complete the cycle. A decrease in available calcium, blockage of the calcium gates, loss of potassium, damage from an infarction, and exposure to certain substances such as ethanol, haloaxanes, and cobalt can each cause a weakening and loss of this function. This condition can also result from damage to the ANS.

Starling's Law of the Heart says that the more blood into the heart, the more blood out of the heart. Generally, as the distention, or filling, of the heart with blood increases, there is a corresponding increase in the force of contraction. Anything that interferes with the amount of blood flow into the heart, therefore, affects the force of contraction, which thus decreases blood pressure, causing hypotension.

Other Disorders

Other disorders of the heart that affect the volume and flow of blood include low blood pressure *(hypotension)*, high blood pressure *(hypertension)*, shock, hemorrhage, persistent clotting in the veins *(thrombosis)*, and blockage of large blood vessels by free-floating blood clots *(embolisms)*.

A lack of adequate supply of blood to the heart itself *(ischemia)* can result in fibrillation. Decreased cardiac output can negatively affect all parts of the body, particularly the brain, kidney, liver, and lungs. Each of these organs has systems that also give feedback to the heart and further affect output. Thus, the nervous system, which maintains a physiological balance in the body fluids, is critical to the proper functioning of the heart.

Cardiotoxic Chemicals

The heart is a crucial pump in the transportation from cell to cell of oxygen, nutrients, cell wastes, hormones, and many other substances vital for body homeostasis. A large number of chemicals and other factors affect heart output in a variety of ways.

Aliphatic Alcohols, Aldehydes, and Glycols

Each of the these exogenous substances either decreases the force of cardiac contraction or diminishes the capacity of the system. Their effects can also result in fibrillation and sudden death.

Halogenate Alkanes

The toxicity of low molecular weight hydrocarbons is known to depress the heart rate, contractility, and conduction of impulses of the heart. These include chlorides, trichloroethylene, trichlorofluoromethane, halothane, chloroform, and methoxyfluorane. The effects are generally reversible in a healthy person who does not have a previous history of cardiovascular dysfunction.

Positive Inotropic Agents

Glycosides of digitalis, strophanthin, and oleandria inhibit the transfer of sodium and calcium across the membrane of cardiac muscle cells. Such a blockage leads to premature contraction of the heart. Aconitine, veratrum alkaloids, palytoxin, and a number of other venoms secreted by animals also use these impulse mechanisms of the heart to produce arrhythmia leading to cardiac arrest and death.

Biotoxins and Other Drugs

Antihypertensive drugs such as hydralazine, guanidine, procainamide, lidocaine, phenytoin, and certain antiarrhythmic drugs such as the local anesthetics can cause ventricular fibrillation and cardiac arrest. Certain drugs, such as mimetics and chemotherapeutics, also act on the nervous system and produce secondary toxicity in the vascular structure. Examples include the anthracyclines, fluorouracil, cyclophosphamide, phenothiazine, and butyrophenone. Many of these are capable of speeding up the heart rate, producing tachycardia.

Gaseous Materials

A variety of materials pose an inhalation danger because of their adverse effects on the heart. Common auto exhaust contains carbon monoxide and various nitrogen oxides; some of these are hazardous in their own right and others may be converted (biotransformed) to other harmful forms when mixed with body chemicals such as water. Nitric acid, nitric oxide, and nitrous oxides are each formed in this manner. Oxygen that is either too rich or pressured and ozone are cardiotoxins. Many of the substances that affect the heart directly also affect the respiratory tract, altering the levels of useable oxygen available to the heart itself and, therefore, resulting in myocardial infarctions.

Other Factors and Summary

In addition to exposure from exogenous chemicals and agents, a variety of other factors affect the output and efficiency of the heart. Poor physical conditioning, obesity, poor nutrition, and age can each contribute to the cardiovascular toxicity of hazardous substances. Detection of toxicity requires a complete physical examination and, thus, is recommended for those who have worked with or have been exposed to hazardous materials. Dysfunction of the cardiovascular system is the most prevalent chronic cause of disease and death of humans in our present industrialized society. The exposure to these chemicals needs to be identified and quantified in order to accurately diagnose cardiac involvement.

THE ENDOCRINE SYSTEM

Like the nervous system, most body activities are also controlled by the *endocrine system,* but its mechanism of control is by chemical action and therefore it acts much more slowly. The glands of the endocrine system produce molecules of chemicals called *hormones* and release them into the blood to travel to distant target organs.

Anatomy and Function

The endocrine system operates as a functional system. This means the glands of the endocrine system are not connected the same way that organs of other systems are, but they each secrete hormonal fluids that regulate other structures of the body and in some cases each other. Figure 5–6 illustrates the glands of the endocrine system. In this section we will briefly discuss hormone function and describe the functions of the organs/glands of the endocrine system.

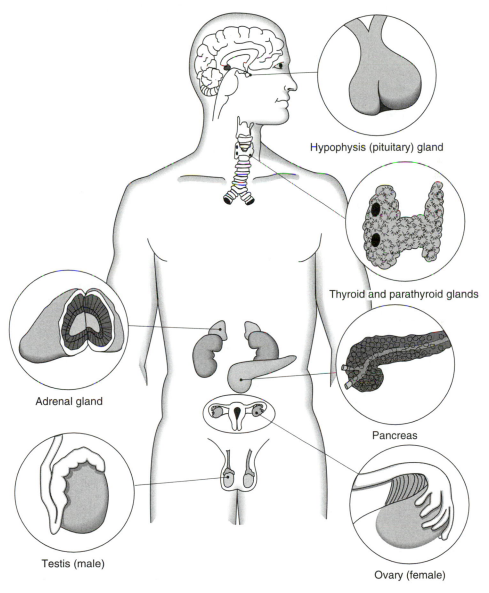

Hypophysis (pituitary) gland

Thyroid and parathyroid glands

Adrenal gland

Pancreas

Testis (male)

Ovary (female)

▶ **FIGURE 5–6**
Locations of the major endocrine organs of the body.

Hormone Function

The term *hormone* is Greek for "arouse," and the function of the endocrine system is to regulate the function and metabolic activity of other cells in the body. Changes in the rate of metabolic processes depend on the specific hormone and the target cell, and typically, one or more of the following changes occur: (a) changes occur in the permeability of the cells' plasma membrane or its electrical state (- vs + charge); (b) enzymes become activated or deactivated; and (c) genetic material itself is stimulated to produce different instructions for increasing or decreasing production of particular enzymes. *Negative feedback mechanisms* occur when rising hormone levels, as the result of stimulus themselves, inhibit further hormone release. Stimuli that activate the endocrine organs may be from other hormones themselves (called *hormonal stimuli),* from changing levels of the formed elements of the blood (called *humoral stimuli),* and in isolated cases from release of *norepinephrine* and *epinephrine* from the nervous system (called *neural stimuli).*

Pituitary Gland

The size of a pea (1 cm^3), the pituitary gland releases *tropic hormones* that stimulate other endocrine glands to secrete their hormones. For this reason it is called the "master" gland of the endocrine system. It also produces several other important hormones:

- *Growth hormone (GH)* is a general metabolic hormone that also has its major effects on the skeletal muscles and long bones of the body. It determines how big and tall we grow.
- *Prolactin (PRL)* is a protein hormone known to target the mammary or milk-producing tissues of the breast.
- *Adrenocorticotropic hormone (ACTH)* regulates the endocrine activity of the cortex of the adrenal gland.
- *Thyrotropic hormone (TH)* is the thyroid-stimulating hormone that influences the growth and activity of the thyroid gland.
- *Gonadotropic hormones* regulate the hormonal growth of the ovaries in females and the testes in the males.
- *Other hormones* include those that trigger ovulation, stimulate release of progesterone and estrogen in females and testosterone in males, each of which also affects growth and development of each gender.

The pituitary also releases *oxytocin* to facilitate contractions during childbirth, *antidiuretic hormones (ADH)* to increase or inhibit urine production, and *vasopressin,* which constricts the small arteries to increase blood pressure.

Thyroid Gland

Located at the anterior base of the throat, the thyroid gland produces several hormones that control basal metabolic rate (BMR), the rate at which glucose is "burned" or oxidized and converted to chemical energy and heat. *Thyroid hormone* is important for normal tissue growth and development, especially in the nervous and reproductive systems of the body. This hormone contains iodine, without which functional hormones of the body cannot be made, a condition leading to goiter. A second hormone is *calcitonin,* which causes calcium to be deposited in the bones and acts antagonistically to the parathyroid hormone produced by the parathyroid glands.

Parathyroid Glands

The parathyroid glands are tiny masses of glandular tissue found on the posterior surface of the thyroid gland. They secrete *parathyroid hormone (PTH),* which is the most important regulator of the balance of calcium ions in the blood. PTH also stimulates the intestine to absorb more calcium and stimulates the kidneys.

Adrenal Glands

These two bean-shaped glands are situated directly over each kidney and produce (a) *corticosteroid,* which is important in regulating mineral content of the blood; (b) *glucocorticoids,* which promote cell metabolism and help the body resist long-term stressors; and (c) *sex hormones,* which are the *androgens* in males and *estrogens* in females. A generalized undersecretion of all the adrenal cortex hormones leads to *Addison's disease,* and an oversecretion can lead to *Cushing's syndrome.*

Adrenal Medulla

When stimulated by perceived danger, the nervous system and the center of the adrenal gland, the *medulla,* release two similar hormones into the bloodstream. *Epinephrine,* also called adrenalin, and *norepinephrine* (noradrenalin) increase the heart rate, blood pressure, blood glucose levels, and dilation of the small passageways of the lungs. The result is more oxygen and increased levels of all other nutrients needed by the brain, muscles, and heart to enable the body to engage its "fight-or-flight" mechanism of self-defense.

Pancreas

Primarily an enzyme-producing organ, the *pancreas* has masses of scattered hormone-producing tissues called the *islet of Langerhans.* Acting like an organ within an organ, these islets produce *insulin* and *glucagon,* both of which help to decrease or increase the amount of glucose (blood sugar) in the blood.

Pineal Gland

Although many chemical substances are found in the *pineal gland,* only the hormone *melatonin* is believed to be secreted in sufficient amounts to be tracked to a significant function: the establishment of the body's day-night cycle and inhibition of the reproductive system so that sexual maturation is prevented until an adult body size has been reached.

Gonads

The female and male *gonads* produce sex hormones that are exactly the same as those produced by the adrenal glands, only in different amounts.

Other Hormone-Producing Tissues

In addition to the major endocrine glands listed above, pockets of cells and tissues throughout the body produce or secrete hormones. These are found in the walls of the stomach, kidneys, heart, and small intestines and are briefly mentioned in other sections of this text. Here we will briefly discuss the remarkable organ of the *placenta,* which is temporarily formed in the uterus of a pregnant woman to provide respiration, nutrition, and excretory functions for the fetus. Unfortunately, it is unable to prevent the transfer of many toxins from mother to the fetus. The placenta also produces hormones such as *human chorionicgonadotropin (HCG)*, estrogen, progesterone, and relaxin, each of which contributes in maintaining the pregnancy.

Endocrine-Specific Toxins

Since the endocrine system itself operates by releasing chemicals into the blood, it isn't difficult to perceive how they can be affected by exogenous chemicals that have also found their way into the blood. Most endocrine glands seem, barring outright malfunction, to operate smoothly throughout life. Exposure to a broad list of exogenous chemicals can lead to malfunction of these glands, and therefore, a host of physical disorders of the body can occur.

The following is a fairly extensive, but not exhaustive, list of chemicals that are known to be toxic to the endocrine system and its hormones: atrozine, metribuzin, nitrofen, trifluralin, benomyl, hexachlorobenzene, tributyl-tin, carbaryl, chlordane, DDT, metabolites, lindane, methoxychlorparathion, toxaphene, cadmium dioxin (2, 3, 7, 8–TCDD), lead, mercury, PCBs, phthalates, and styrenes.

Conclusion

When you compare the organs and glands of the endocrine system with other organs of the body, they may seem small and insignificant. You would need to collect all of such tissue from several adults to total even a pound. Bits and pieces of this system are found throughout the body, and thus it lacks the structural or anatomical continuity typical of most other organ systems. Functionally, however, the endocrine organs are very impressive and they play a major role in maintaining body homeostasis.

THE NERVOUS SYSTEM

The nervous system is the command, control, and communication system of the body; it is the master controller of every activity of the body. Much like a computer network, the central nervous system communicates with rapid chemical impulses that cause almost immediate and specific responses.

The nervous system is very complex and is comprised of the *central nervous system (CNS)* and the *peripheral nervous system (PNS)* (Figure 5–7). The CNS is, basically, the brain and spinal cord, whereas the PNS consists of the portion that extends from the CNS to the individual sensors and motor units throughout the body. All that we know, sense, experience, and remember about the world is conveyed to us by specialized nerve cells called *neurons*. These special cells control the circulation of blood, respiration of air, digestion of food, and voluntary movement by muscle control. These cells also allow us to use our senses of smell, touch, taste, sight, and hearing. Furthermore, these cells allow us to think and to remember.

Anatomy and Physiology

Neurons have a unique communication ability in the body, all the more amazing because they do not make actual contact with one another. There are numerous "breaks" or "gaps" along the pathways on which nerve messages, or "impulses," travel. These gaps are called *synapses,* and the impulse is chemically transmitted

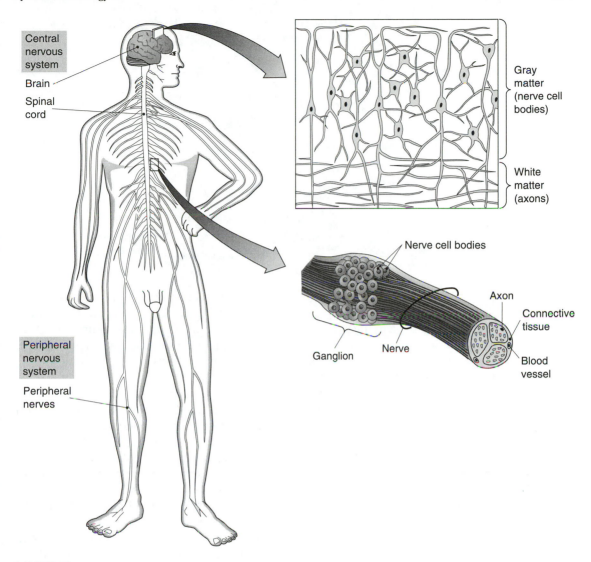

▶ FIGURE 5–7
The nervous system.

across them (Figure 5–8). This process is facilitated by chemicals, neurohumoral transmitters, which are released into these gaps from the presynaptic terminals. The neurohumoral transmitters diffuse across the small spaces, one to two hundred angstroms, to receptors located in the membranes of the neurons on the opposite side, postsynaptic receptors. Different neuronal processes, neurons, or groups of neurons respond to different neurohumoral transmitters. Therefore, different chemicals, or substances, have different effects on the nervous system.

Because neurons are so specialized, they have lost the ability to care for themselves. Therefore, a specialized population of nurse or "glial" (glue) cells are required. These glial cells are a very important part of the nervous system; they surround,

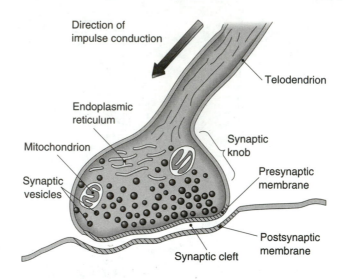

▶ **FIGURE 5–8**
The structure of a synapse.

protect, and facilitate the activities of the neurons. Glial cell populations include the Schwann cells, the microglia, the astrocytes, the oligodendrocytes, and the ependymal cells. Damage to the glial cell populations can be disruptive to neuron function or lethal. Since neurons cannot divide and repopulate, neuronal death is permanent. When you are born, you have the most neurons you will ever have; subsequently, the neuronal population decreases throughout life.

Neuronal cells have a nerve cell body and one or more appendages or processes. Every nerve cell has one process called the *axon*. All information leaves the neuron via the axon. The axon may connect to a sensory receptor, other neurons, or to motor (muscle) units. Additionally, neurons may have no, or up to 50,000 other processes called *dendrites*. Dendrites are responsible for receiving information and conveying it toward the neuron cell body. Throughout life neurons may establish new connections and make new synapses. Such new connections contribute to learning and memory. Thus, even though we lose neurons, in absolute number, throughout life, we make new connections with the ones that remain and can learn new tricks.

The cell process may extend long distances, over a meter in some cases, to bridge information from one part of the body to another. In the CNS like groups of these processes travel together to form "tracts." Groups of the nerve cell bodies frequently conglomerate in what are called "ganglia" or "nuclei." When groups of sensory and motor fibers travel together, as is common in the PNS, they are typically covered by a layer of tough connective tissue and called "nerves."

Neuronal Populations

The neuronal population may be divided by function. We have:

Sensory Cells

Sensory cells are unipolar cells that have their cell bodies safely tucked away in groups called ganglia. Most of these are located close to the CNS, although a few are located in the periphery. The function of these cells is to sense hot, cold, pain, pressure, touch, sight, vision, sound, balance, and all other information we receive regarding our internal and external environments.

Motor Cells

Motor cells relay information out of the CNS. It tells our muscles when to move and our glands when to secrete; it relays all of our conscious and unconscious thoughts for execution. Those actions that we explicitly and voluntarily control are relayed through the voluntary nervous system. Those actions we do not typically control and, probably, are better off not worrying about (respiratory rate, heart beat, blood pressure, and digestion) are nominally under the control of the autonomic nervous system (ANS). Because neurons can have only one set of actions, control requires two sets of neurons, one to stimulate and one to relax. The sympathetic nervous system and the parasympathetic nervous system perform these opposite functions for the ANS.

Intermediate or Internuncial Cells

Internuncial cells are located throughout the CNS and act as an interface between sensory and motor neurons. They are particularly involved in reflex arcs and actions.

Association Cells

Association cells are very abundant in the cerebral cortex of humans. Groups of these cells are found between input and output of thoughts and information. The ability of these cells to communicate to each other and have "discussions" provides us with many of our unique abilities to think, rationalize, and be self-aware.

Glial Populations

Schwann Cells and Oligodendrocytes

These cells tightly, and repeatedly, wrap themselves around some of the nerve cell processes to form what is termed *myelin*. Thus, nerve fibers may be either wrapped in a myelinated sheath (white matter) or not myelinated (gray matter, or fibers that have no sheath). Myelinated fibers tend to be high speed, which is crucial to, for example, motor neurons. Loss of myelination causes a dysfunction of these cells. Multiple sclerosis is a degenerative disease that destroys the myelin sheath causing a progressive loss of motor function.

Astrocytes

The principal function of the astrocytes is to maintain the blood-brain barrier. These cells interpose themselves between the capillaries of the circulatory system and the neurons of the nervous system. As such, they are positioned to control the flow of fluid from the circulatory system into the CNS. This control of the interstitial environment is known as the blood-brain barrier. Unfortunately, like the placenta, this barrier is not impervious to all toxins, such that many poisons do pass from blood to brain.

Microglia

The microglia are small, mobile, phagocytic cells. Because the blood-brain barrier prevents the free movement of the formed elements into the CNS, the functions normally associated with the WBC need to be picked up, thus the need for the microglia.

Ependymal Cells

The CNS is perfused by a special fluid called the cerebral spinal fluid (CSF). The ependymal cells filter and process the blood plasma to produce CSF. Because the blood-brain barrier prevents the free flow of plasma into the interstitial spaces and lymphatics do not typically permeate the CNS, these functions are assumed by the circulation of CSF.

Toxic Damage

The nervous system has the ability to tolerate and adapt to certain damages. Therefore, it can continue to function until the reserve capacity of the system is exceeded. Toxic damage, thus, may be quite extensive before it becomes manifest. Initial toxic damage can be detected only by extensive testing. The earliest, and most sensitive, signs of damage include alterations in gait and motor/visual performance, irrational thought, and change in emotional status. (Alcohol intoxication is an example of toxic damage.) Changes due to toxic insult can be divided into three categories, depending on the type of damage sustained:

Sensory Alterations

These alterations include damages to any of the five basic senses: sight (covered in section on eye), hearing, taste, smell, and those of touching, including pain, pressure and temperature. Damage may occur in any location along the sensory pathway up to the sensory receiving areas. Examples include heavy metal poisoning, such as lead and mercury, resulting in deafness and loss of sight. A variety of inorganic salts and organophosphorus compounds lead to a loss of sensory functions.

Motor Alterations

Neuronal damage and disruption produce motor dysfunction. Damage to motor nerve axons or terminals results in weakness and, often, paralysis of the involved muscles. An example of such damage is observed after absorption of isonicotinic hydrazide.

Integrative Alterations

Learning is retention and memory of the functions of symbol formation (reading), sensory-motor integration (coordination), and emotional state. These are all integrated together by associative neurons that are susceptible, collectively and individually, to toxins. Changes in emotional states are manifested by apprehension, irritability, and lability in persons exposed, for example, to low levels of inorganic mercury. Absorption of low doses of carbon monoxide results in depression and memory loss.

Types of Toxicants

Toxicants may be classified according to their primary toxic action. Classification by effect provides a framework for the discussion of toxic agents.

Anoxia

Anoxic damage can result from inadequate oxygen supply to the cells. This may be due to neurohumoral transmission, blocking by agents, inadequate blood flow, or by interference with cell metabolism. Examples are barbiturates, carbon monoxide, cyanide, azide, and nitrogen trichloride.

Demyelination

Many axons within the nervous system are wrapped with the protective and impulse-facilitating myelin sheath. Agents that selectively damage these coverings disrupt or interrupt the conduction of high-speed neuronal impulses. Examples of such agents are isonicotinic acid, hydrazide, triethyltin, hexachlorophene, lead, thallium, and tellurium.

Muscular Atrophy

Damage to peripheral motor nerves, or degeneration of axons following demyelination, results in muscle atrophy. The following group of toxic compounds causes this "dying back" neuromuscular symptom: alcohol, acrylamide, bromophenylacetylurea, carbon disulfide, hexanedione, and organophosphorus compounds.

Neuronal Dysfunction

Agents that have, as their primary cause of damage, cell body, perikaryon degradation result in degradation of neuronal cell function. Such damage to the nerve cell body diminishes the ability to, for example, synthesize protein for the normal function of the entire neuron. Examples of such agents include organomercury compounds, vinca alkaloids, iminodipropionitrile, and vinblastine.

Synaptic Dysfunction

The synaptic clefts and synapses are particularly vulnerable and susceptible to a wide variety of chemicals. Damage, or effects, may be either reversible or irreversible. Many of the wartime nerve gases were specifically designed to irreversibly bind with receptors, causing the rapid death of the organism. These lethal preparations are not in general use today. However, a wide variety of less lethal analogues of these agents are commonly used in agriculture and industry. As the neurohumoral receptor sites are intended to respond to chemical transmitters, it is not surprising that these sites frequently are involved in the severe poisoning of humans. Examples include many natural toxins such as botulinus toxin, tetrodotoxin, and saxitoxin. Industrial toxins include the organophosphate and organochlorine insecticides, DDT, pyrethrin and lead.

CNS Lesions

A final group of neurotoxins are those that may cause localized CNS lesions. Areas of the CNS that lack the blood-brain barrier are especially susceptible. Since the blood-brain barrier is imperfect, many common drugs, such as nicotine, cocaine, hallucinogenics, heroin, and alcohol, readily permeate it. Alcohol intoxication leads to ataxia, rigidity, tremor, facial grimacing, and both emotional instability and dementia. Examples of other agents include methione sulfoximine, glutamate, gold, thioglucose, acetylpyridine, trimethyltin, pyrithiamine, DDT, mercury, and manganese.

Blood-Brain Barrier

Fortunately, not all chemicals and substances that are exogenous to the body can target and reach the nervous system. As discussed in the section on glial cells, the body has a number of mechanisms that wall off and protect the CNS vis-a-vis the blood-brain barrier. A number of pharmacological agents such as analgesics, anesthetics, and tranquilizers readily penetrate the barrier because they are nonpolar and lipid-soluble. However, highly polar and nonsoluble compounds are unable to permeate this barrier because of the size of their molecules or their inability to fit into transport mechanisms. Physiologically, this means that the capillaries in the CNS are the least permeable capillaries in the entire body. Unfortunately, treatment of certain conditions, for example, bacterial infections of the CNS, are difficult because it may be impossible to introduce therapeutic agents into the CNS through the normal routes of drug administration.

THE REPRODUCTIVE SYSTEM

For the purpose of this section, we will define reproductive toxicity as a dysfunction of any stage in the reproductive process induced by chemical, biological, or physical agents. The survival and perpetuation of any species depends on its ability to generate new life. Unfortunately, because of the slow onset of many reproductive effects, the potential chemical effect on reproductive function is an all too frequently misunderstood and ignored toxicologic phenomenon. Figures 5–9 and 5–10 illustrate the male and female reproductive systems.

Anatomy and Physiology

Gametes are haploid germ cells produced by both the male and female gonads. (*Haploid cells* have only half of the chromosomes of normal cells, so that when two gametes unite, the full complement of chromosomes are present.) The primary sex organs are called the testes in the male and the ovaries in the female. In the male, sperm is produced in the seminiferous tubules (Figure 5–11). The sperm subsequently travels along a series of ducts until it exits at the distal end of the penis. Along this tract are several accessory glands that contribute secretions for neutralizing the acidic environment of the female vagina and fructose for nourishing the sperm on their journey.

▶ **FIGURE 5–9**
The male reproductive system.

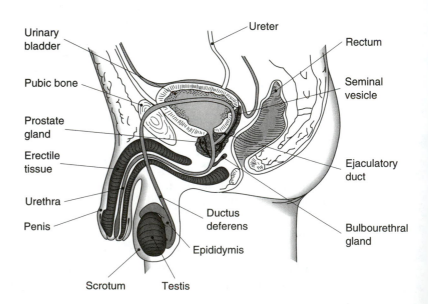

Urinary bladder

Ureter

Rectum

Pubic bone

Seminal vesicle

Prostate gland

Erectile tissue

Ejaculatory duct

Urethra

Penis

Ductus deferens

Bulbourethral gland

Epididymis

Scrotum   Testis

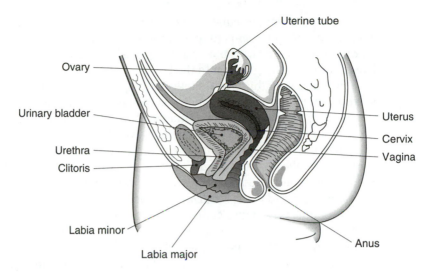

▶ **FIGURE 5–10**
The female reproductive system.

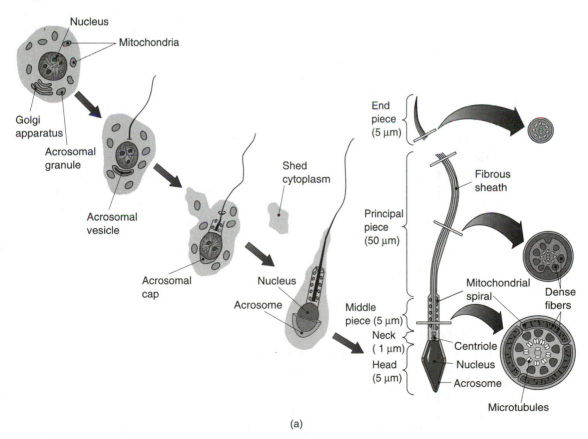

(a)

▶ **FIGURE 5–11**
Spermiogenesis and spermatozoon.

For procreation, ejaculation needs to occur within the vagina of the female. The sperm swim up the vagina to the uterus. At the opposite end of the uterus, a pair of fallopian tubes attach. The ovum, or egg, which has matured in the ovaries, is released and, ideally, sucked into the fallopian tubes.

The success of this process relies on a sperm being able to penetrate the corona surrounding the ovum. Each sperm is capped by an acrosome containing enzymes designed to eat away at the corona surrounding the ovum. If sufficient sperm are present, the corona is gradually etched away and one sperm penetrates the ovum. This meeting of the sperm with the ovum results in conception. This process is extremely complicated, and its success depends on a variety of mechanisms producing fluids, hormones, and activity within the reproductive system. Figure 5–12 depicts the stages involved in the fertilization process.

The gametes contain 23 chromosomes in the female and 23 in the male. The sum of these is the normal complement of 46 chromosomes in the human. The mixing and matching of the male and female sets of chromosomes provide the genetic instructions to form the characteristics of the developing human being.

The now-fertilized ovum, called a *zygote,* is the beginning of a new human being. The zygote continues its journey along the fallopian tube to reach the *uterus.* The zygote has divided numerous times during its journey and is now a small ball of tightly packed cells. It implants into the wall of the uterus as an *embryo.* A complicated structure, called the *placenta,* develops and transmits nourishment from the mother. The embryo is attached to the placenta through the *umbilical cord* (Figure 5–13). Exchange of all fluids and substances between the mother and embryo are made through this placenta. This placental barrier was at one time thought to protect the embryo from substances and toxins found in the mother. While the placental barrier is quite good, like the blood-brain barrier, it can be penetrated by a wide variety of toxins, such as nicotine, caffeine, cocaine, and heroin.

For the first couple of months of development, the embryo is called an embryo. After the ninth week, it is referred to as a fetus. The development of the embryo and fetus is very precise and staged. Any interruption of this general development also disrupts particular development. Early interruption usually ends in spontaneous abortion. Subsequent interruption depends on a variety of factors. The drug thalidomide, for example, was marketed from 1957 to 1961 to be used by pregnant women to control morning sickness; however, it interrupted limb bud formation and resulted in missing or reduced arm and leg formation in nearly 8,000 children.

Reproductive toxicity refers to chemically produced damage to any of the reproductive fluids, ducts or tubules, accessory organs, external genitalia, or to the gamete, zygote, embryo, fetus or its mechanism for nutrition.

Targets for Chemical Toxicity

Research has identified several vulnerable sites within the reproductive system where chemicals can cause dysfunction.

Cytotoxicity

Cytotoxicity refers to chemicals that are capable of reducing the number of sperm or ova. These include the anticancer drugs of alkaloids, alkylating agents, antimetabolites, and antitumor antibiotics. Mutagens may also be included under this classification of chemicals.

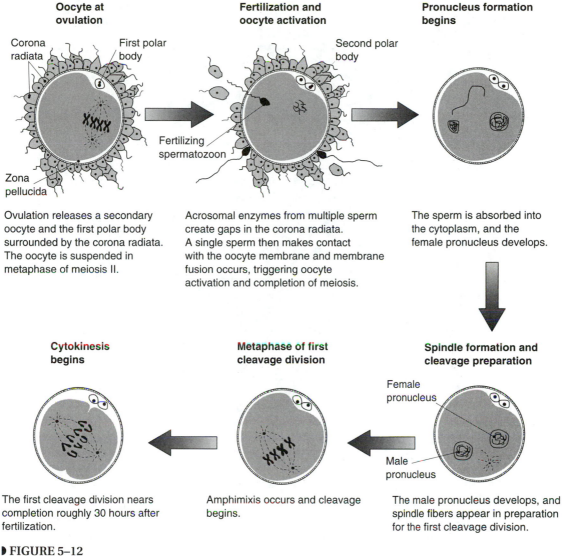

Oocyte at ovulation

Corona radiata   First polar body

Fertilizing spermatozoon

Zona pellucida

Ovulation releases a secondary oocyte and the first polar body surrounded by the corona radiata. The oocyte is suspended in metaphase of meiosis II.

Fertilization and oocyte activation

Second polar body

Acrosomal enzymes from multiple sperm create gaps in the corona radiata. A single sperm then makes contact with the oocyte membrane and membrane fusion occurs, triggering oocyte activation and completion of meiosis.

Pronucleus formation begins

The sperm is absorbed into the cytoplasm, and the female pronucleus develops.

Cytokinesis begins

The first cleavage division nears completion roughly 30 hours after fertilization.

Metaphase of first cleavage division

Amphimixis occurs and cleavage begins.

Spindle formation and cleavage preparation

Female pronucleus

Male pronucleus

The male pronucleus develops, and spindle fibers appear in preparation for the first cleavage division.

◗ **FIGURE 5–12**
Fertilization.

Nerve Transmission

Neural transmission is critical to a number of reproductive functions. Impotence and lack of sex drive is a symptom of alcohol intoxication. Some drugs, such as alcohol, can stop sexual arousal before it can get started.

Hormone Action

Timing is critical to the reproductive process. The endocrine system produces the hormones and orchestrates the process. This is a very complicated and multifaceted process, much more so than the ease of pregnancy in our society might suggest. Unfortunately, many of the problems, especially those related to birth defects, are detected only by large epidemiological studies long after the initial damage has occurred.

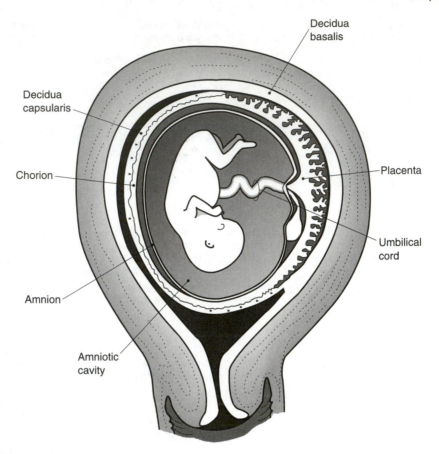

▶ **FIGURE 5–13**
Growth of fetus and uterus.

Peptidal and steroidal hormonal action depends on numerous steps, any of which can affect the reproductive process. Methyl-p-tyrosine, for example, blocks the synthesis of norepinephrine; methyldopa biotransforms and competes for binding sites and displaces norepinephrine; imipramine blocks transmission; phenylephrine and isoproterenol mimic; phenoxybenzamine and pargyline interfere with enzymatic breakdown and assimilation. Hormones administered to livestock work their way up the food chain. Diethylstilbestrol, used to increase white meat in chickens, has been linked with gynecomastia (breast enlargement) in young boys in developing countries.

Fluid Milieu

The volume and composition of semen and vaginal fluids are critical to the fertilization process. The environment available to the developing embryo requires meticulous maintenance. These fluids may be altered by diuretic drugs and environmental chemicals.

Other

A variety of other factors may influence the process. Though clearly not exhaustive, the following is a list of some such dysfunctions:

- Sertoli cell formation
- Sperm enzyme viability
- Ovulation
- Fertilization
- Early pregnancy
- Implantation
- Sexually transmitted disease
- Epididymal function
- Sperm maturation
- Oogenesis
- Folliculogenesis
- Corpus luteal formation and function
- Sexual behavior

Human Risk

Humans are exposed to a large number of agents that are hazardous to their reproductive capacity. These are commonly found in our environment. The list includes steroid hormones, chemotherapeutic agents, metals and trace elements, pesticides, food additives and contaminants, industrial chemicals, radioactivity, and consumer products. The impact and effect these have on any particular pregnancy is highly dependent on the individuals involved. Given the uncertainties involved, careful planning for parenthood is essential to ensuring the well-being of the progeny. Paradoxically, the period in which the embryo is most sensitive to toxins is the first month when many mothers do not realize that they are pregnant.

SUMMARY

In this chapter we have introduced readers to the general concept of systemic toxicology and have reviewed a number of organ systems:

- The *integumentary system,* or *skin,* has the responsibility as the first line of defense for the body. The skin protects the body, keeping water and other precious nutrients in the body and also keeping unwanted things out. Foreign substances cause a host of problems, including infections and allergies, burns, irritations, corrosion, photosensitivity, chloracne, and other skin reactive patterns.
- The *blood* is a complex connective tissue that serves as the transportation fluid for oxygen, nutrients, hormones, and other substances and is referred to as "the river of life." Plasma is the liquid portion, and the formed elements are the red blood corpuscles, WBCs, and thrombocytes. Substances that result in harmful effects on the blood are called hematoxins.
- The *cardiovascular system* is the pump and the system of tubes that provides the blood to various parts of the body. The main organ is the heart. Substances that are harmful are called cardiotoxins, and their effects may alter the impulse rate,

disturb the site of the impulse origin, compromise the strength of the heart's contractions, and cause other disorders that can affect the volume and flow of blood.

▶ The *endocrine system* helps to control the functions of the body by producing chemicals, or hormones, and circulating them throughout the body. The principal organs or glands of this system are the (a) pituitary or "master" gland, (b) thyroid gland, (c) parathyroid gland, (d) adrenal glands, (e) medulla, (f) pancreas, (g) pineal glands, (h) gonads, and (i) several other tissues located specifically within the stomach, kidneys, heart, and small intestine. The placenta also produces and distributes hormones in the management of human pregnancy.

▶ The *nervous system* consists of both the central nervous system and the peripheral nervous system. This system controls the functions of the body by chemical flux impulses that are sent along pathways called neurons and across many junctions of these neurons called synapses. The impulses pass across these synapses by a chemical process that can be affected by harmful chemicals called neurotoxins.

▶ The *reproductive system* is affected by a large number of hazardous substances called reprotoxins. These toxins especially affect the gonads, which include the ovaries in the female and the testes in the male. From development of the gamete, to sperm count and fertilization, and to development of the embryo, especially during the first three months, there is a large risk from exposure to hazardous substances, viruses, and bacteria.

The discussion of systemic toxicology continues in Chapter 6.

STUDY QUESTIONS

1. Define *systemic toxicology* and identify its relationship to target organs.
2. Briefly discuss each of the following concepts and components of systemic toxicity: hypoxia, anemia, cardiotoxins, hematoxins, neurotoxins, and reprotoxins.
3. What is the purpose and the function of the integumentary system? Name three chemicals or chemical groups that adversely affect this system.
4. What is the purpose and the function of the blood? Name three chemicals or chemical groups that adversely affect this system.
5. What is the purpose and the function of the cardiovascular system? Name three chemicals or chemical groups that adversely affect this system.
6. What is the purpose and the function of the endocrine system? Name three chemicals or chemical groups that adversely affect this system.
7. What is the purpose and the function of the central nervous system? Name three chemicals or chemical groups that adversely affect this system.
8. What is the purpose and the function of the reproductive system? Name three chemicals or chemical groups that adversely affect this system.

6

Systemic Toxicology-2

COMPETENCY STATEMENTS

This chapter provides a brief introduction to the principles of systemic toxicology. After reading and studying this chapter, students should be able to meet the following objectives:

▶ Describe the purpose and the function of the respiratory system and its relationship to systemic toxicology.

▶ Describe the purpose and the function of the kidney and its relationship to systemic toxicology.

▶ Describe the purpose and the function of the liver and its relationship to systemic toxicology.

▶ Describe the purpose and the function of the eye and its relationship to systemic toxicology.

▶ Describe the purpose and the function of the immune system and its relationship to systemic toxicology.

INTRODUCTION

In Chapter 5 we laid the groundwork for our study of systemic toxicology and discussed the following target organ systems:

▶ Integumentary system (skin)

‣ Blood
‣ Cardiovascular system
‣ Endocrine system
‣ Central nervous system (CNS)
‣ Reproductive system

In this chapter we continue the discussion of systemic toxicology with the following organs and organ systems:

‣ Respiratory system
‣ Liver
‣ Kidney
‣ Eye
‣ Immune system

THE RESPIRATORY SYSTEM

Exposure to toxicants via inhalation occurs in all phases of human activity. Breathing is an autonomic function, so even if you tried to stop breathing, you would eventually pass out and your body would begin breathing again on its own.

Absorption of harmful airborne substances in the workplace is the most important method of exposure for workers and, therefore, is of great concern for industrial toxicologists. Many industrial gases and particles suspended in the atmosphere are capable of damaging the respiratory system because they cause inefficient or suppressed respiratory function.

System Function

The trillions of cells in the body require an abundant and continuous supply of oxygen to carry out cellular respiration and energy production. The by-product of these activities is carbon dioxide. The lungs provide a means for the exchange of oxygen and carbon dioxide between the atmosphere and blood.

System Anatomy

Many textbooks break this system into the upper and lower respiratory systems, but in this chapter we will merely follow the air as it enters the nose and proceeds through the system to be absorbed by the blood. Figure 6–1 illustrates the structures of the respiratory system.

1. *Nose:* The only external portion of the respiratory system is the nose. As air enters, it is moistened and warmed by blood-rich connective tissue in the *mucosa* lining. In addition, sticky *mucus* is produced by this lining, which traps large dust particles, incoming bacteria, and other debris. The sense of smell also originates in the nose and contributes to our pleasure and serves as a warning system of the body.
2. *Pharynx:* The pharynx is a muscular passageway about 12.7 centimeters (5 inches) long that directs incoming air toward the larynx. Food also passes through this airway but is separated and continues directly into the *esophagus,* or tube leading to the stomach.

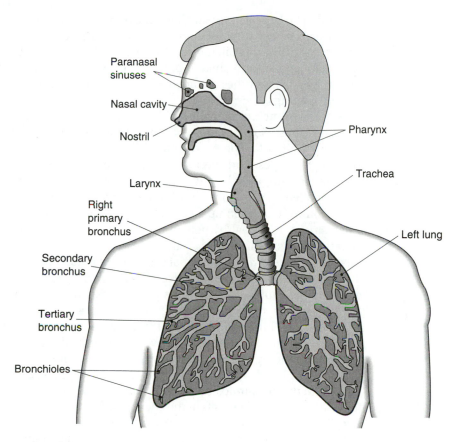

▶ FIGURE 6–1
Structures of the respiratory system.

3. *Larynx:* The larynx, or voice box, is a tube that allows air to pass by the esophagus and routes it into the lungs. The *epiglottis* is regarded as the "guardian of the airways" and is a spoon-shaped flap that covers the trachea whenever we swallow food or fluids by forming a lid over the opening of the larynx and routes food into the esophagus as discussed above. If anything enters the larynx other than air, a *cough reflex* expels the substance and prevents it from continuing to the lungs.

4. *Trachea:* The trachea is known as the "wind pipe" and is a tube that guides air from the larynx to the bronchus. This tube is resistant to collapse and is held open by C-shaped rings of cartilage.

The nose, pharynx, larynx, and trachea are lined with a mucous membrane that traps particles and also dissolves substances and makes some of them available to the tongue for the taste response. This membrane is lined with *ciliated mucosa,* the cilia of which beat continuously and in opposite direction from the incoming air. The mucosa propel mucus, loaded with debris, away from the lungs and back up to the throat where it can be swallowed. This action is called the *mucociliary escalator,* and without these cilia, coughing is the only means of preventing mucus from accumulating in the lungs.

5. *Bronchus:* The largest, right and left, tubes of the lung begin at the first division of the trachea and constitute the primary bronchi. These then divide into secondary bronchi, again into bronchioles, and then each continues to divide again and again into smaller and smaller branches called *respiratory and terminal bronchioles.* All of these tubes are lined with smooth muscle, which controls the size of their openings.

6. *Alveoli:* The bronchioles finally divide into even smaller ducts called the *alveolar ducts,* and these ducts eventually connect into clusters of small, thin-walled air sacs called *alveoli.* There are millions of clustered alveoli, which resemble bunches of grapes, and these make up the bulk of the lungs. The walls of the alveolus are composed largely of a single layer of epithelial cells, and the surface is surrounded with a "cobweb" of pulmonary capillaries. It has been estimated that if you were to open up all of the alveolar sacs and lay them side by side on a flat surface, in a healthy person they would cover between 70 and 80 square meters, or the size of a racquetball court.

7. *Respiratory membrane:* The actual exchange of atmospheric gases occurs across the respiratory membrane. The alveoli cell walls are on one side with gas flowing past, and the capillaries are on the other side with blood flowing past. The two are fused together to form the respiratory membrane.

8. *The lungs:* The lungs are the fairly large organs that contain the "upside down tree"-like structure of airways we have just described. There are three medium-size lobes in the right side of the chest and just two lobes on the left side to make room for the heart within the thoracic cavity. The lungs are surrounded by the *pleural cavity,* the surface of which is covered with a slippery membrane called the *visceral pleura.* The walls of the chest cavity are lined with the *parietal pleura.* These pleural membranes produce a slippery secretion called *pleural fluid* that permits breathing movements of inhalation and exhalation; as the lungs expand and contract, these membranes slide across one another and are lubricated by the pleural fluid. *Pleurisy* is an inflammation of the pleura caused by a decrease in pleural fluid. The result is an intense stabbing pain in the chest with each breath.

Physiology

The respiratory tract is unique in that its vital functional elements are continuously exposed to the external environment. Fortunately, a series of mechanisms warm, moisten, and cleanse the air that is inhaled. Therefore, all particulate matter that is inhaled is not necessarily absorbed. Protective mechanisms begin with the mucosa of the nose and continue throughout the entire respiratory tract. The final line of defense for the respiratory system is the macrophages found in the alveoli. These hungry cells wander in and out of the alveoli picking up particles of foreign substances, carbon, wastes, and other debris.

It is important to remember that all gas exchanges occur according to the principle of diffusion. Molecules of oxygen flow from inhaled air, across the alveolar membrane, and into the blood where they are then absorbed into the hemoglobin molecules. Carbon dioxide dissolved in the plasma, on the other hand, flows from the capillaries, crosses the membranes, and then is carried away with the exhaled breath. The effectiveness of this mechanism depends on several factors, principally on the solubility of the substance and its ability to bind with proteins in the blood.

The lung, itself, is also able to biotransform and metabolize many substances as well as facilitate their removal by exhalation.

Additionally, all foreign substances inhaled do not necessarily reach the alveoli, nor are all those that reach the respiratory membrane transported across to the bloodstream. Some fraction of the substances will be immediately exhaled; another portion will be diluted within the tubes, dissolve in the mucus of the upper respiratory tract, or be coughed, sneezed, or otherwise forcefully exhaled and returned to the exterior environment. Some substances will be "taken out" by the immune system's macrophages. However, much of the gases and vapors will pass directly into the bloodstream, which is why the respiratory system represents the principal pathway for absorption in the industry environment.

Respiratory Disorders

Because the respiratory system is particularly vulnerable to infections, we will briefly discuss the group of diseases collectively referred to as *chronic obstructive pulmonary disease (COPD)* and *lung cancer.* Exemplified by *bronchitis* and *emphysema,* COPD is a major cause of death and disability in the United States. COPD victims have the following conditions or features in common: (a) almost every patient has a history of smoking tobacco; (b) labored breathing, called *dyspnea,* occurs and becomes more severe over time; (c) infections and chronic coughing are common; (d) victims retain carbon dioxide, become hypoxic (oxygen-poor blood), and have respiratory acidosis (low-pH blood), and (e) victims ultimately develop respiratory failure.

Emphysema

Emphysema is a condition in which the lungs lose epithelium to become brittle and less elastic due to chronic inflammation: the airways collapse during expiration and obstruct air flow. These patients must exhale an incredible amount of air to empty their overinflated lungs, and the condition leads to a permanently expanded barrel chest.

Bronchitis

In this condition the mucosa of the lower respiratory passages become severely inflamed and produce excessive amounts of mucus. If this extra fluid in the lungs builds up, it can result in *pneumonia* and a subsequent loss of gas exchange capacity. It dramatically increases the risk of even more serious lung infections. Chronic bronchitis patients often turn blue *(cyanosis)* due to hypoxia and carbon dioxide retention in the early stages of the disease.

Response to Toxicants

Foreign substances are classified according to how they affect the respiratory tract and are termed "respiratory toxins" whenever these effects are harmful. The following paragraphs explain some of these effects and the toxins that cause them.

Direct Airway Irritation

The tone of the smooth muscle lining the respiratory tree is immediately influenced by inhalation of chemicals and aerosols such as carbachol or acetylcholine, which causes constriction. Dilation of the smooth muscle is produced by inhalation of chemicals such as isoproterenol. Ammonia and chlorine are classic examples of *irritant gases* that cause such symptoms. Arsenic compounds and hydrogen fluoride are also known to cause constriction of the airways. Additionally, the cellular response to irritation is to allow fluid to accumulate within the cells in a condition known as *edema*. Damage to the cell membranes, swelling, and other inflammation further constrict the tubes, the cells are weakened, and they fall easy prey to bacteria and other infections of the airway.

Cellular Damage

A variety of materials produce damage to cells of the airway tubes and alveoli. This damage leads to an increased permeability of the respiratory membrane and a "weeping" of fluid into the external spaces. Ozone and nitrogen dioxide are common examples of such necrosis producers in our environment. Phosgene, cadmium oxide, nickel compounds such as carbonyl, and hydrochloric acid are found in the industrial environment.

The net effect of fluid accumulation is that the lung is prevented from a free exchange of atmospheric gases. The effect is much like drowning and is referred to as chemical asphyxiation. (Simple asphyxiation is the lack of oxygen in the atmosphere being breathed, or less than 19.5% oxygen in the air we breathe.)

Production of Fibrosis

The term *fibrosis,* as it is used here, refers to the buildup of scar tissue, usually from replacement of destroyed tissue by fibrous connective tissue, and can result in the blockage of airways. Fibrous and scar tissues do not allow the free exchange of gases as do the normal cells of the alveolus; therefore, the proliferation of these tissues occurs at the expense of normal alveolar tissue and results in decreased lung capacity. Examples of such *pneumotoxic* materials include beryllium, coal dust, the silicates, and asbestos, the last two of which will be discussed in the following sections.

Silicosis   This is a progressive lung disease brought on by the inhalation of rock dusts. Quartz is the most common form which, when heated, produces tridymite or crystobalite, each of which is notably potent in producing fibrosis of the lung. Coal Miner's Disease, also known as "black lung," is one of the earliest diagnosed forms of silicosis and occurred as the result of breathing coal dust.

Asbestoses   This term identifies a large group of hydrated silicates which, when milled or crushed, separate into minute fibers. These fibers are further identified as chrysotile, amosite, crocidolite, anthophyllite, tremolite, and actinolite. Recognized very early as causing a special form of respiratory disease, asbestoses are made even more active by the effects of tobacco smoke. Asbestoses have led to the development of OSHA standards that regulate the levels of dust in the workplace for asbestos and for other airborne particulates.

Induction of Allergic Response

Numerous materials, when inhaled, may cause reactions such as chronic coughing, sneezing, production of mucus, tearing of the glands surrounding the eye, and constriction of the pulmonary tracts. Examples of these phenomena include Farmer's

Lung from the inhalation of organic dusts (dust from chicken manure); mushroom picker's lung; and diseases from sugar cane, maple bark, cheese spores (penicillin), moldy sawdust, cotton flax, and hemp dust. Synthetic agents that produce an allergic type of response include toluene disocyanate (TDI), methylisocyanate, and iron oxides.

Production of Pulmonary Cancer

Tobacco smoke is primarily a source of irritation and death to the ciliated epithelium located in the area of the respiratory tree. Cigarette smoke also decreases gas exchange, creates carbon monoxide, which preferentially binds to hemoglobin in the blood, and interferes with alveolar macrophages (antigen killers). Many of the over 2,000 chemicals released by burning tobacco can lead to cancer of the lung and the respiratory tract. Other notable carcinogens of the lung include nickel compounds, benzopyrene from coke oven emissions, and chromate salts.

Induced Oxygen Deprivation

Classified as asphyxiants are a large number of gases that deprive the body tissues of much needed oxygen. Essentially, these inert gases displace oxygen, which leads to suffocation. Such gases include nitrogen, helium, methane, neon, and argon. Other chemical asphyxiants prevent the tissues from getting enough oxygen and include carbon monoxide and cyanide, which deactivates the electron transport enzymes while the former essentially prevents hemoglobin from transporting oxygen molecules. Normal air contains approximately 21% oxygen; the minimum levels necessary to sustain life are 19.5%. Acute asphyxiation or suffocation can occur when gaseous molecules have a vapor density heavier than air and thus displace the oxygen in the breathable environment at head level.

Conclusion

The high incidence of pulmonary disease and lung cancer in the United States has directed much attention to the toxic materials in the air we breathe, especially in the occupational settings. Therefore, it follows that pulmonary (respiratory) protection programs are among the most essential of any industrial safety and hygiene program, popularly called a Health & Safety Program (HASP).

THE LIVER

The study of the effects of chemicals on the liver began at the turn of the century. The evidence is now quite clear that this organ, which is essential in detoxifying and biotransforming toxic substances, is quite vulnerable to toxic substances itself.

The liver is the largest gland in the body and comprises 3% to 4% of total body weight. It is a soft organ, albeit solid, that is divided into four lobes and is located in the upper right abdominal quadrant of the body overlying the stomach. Oxygenated blood is delivered to the liver by the *hepatic artery,* and blood from the stomach, pancreas, and spleen is delivered by the *hepatic portal vein* on its way back to the heart for recirculation. Although this portal blood is low in oxygen, it contains high concentrations of dissolved nutrients and virtually all of the substances absorbed from

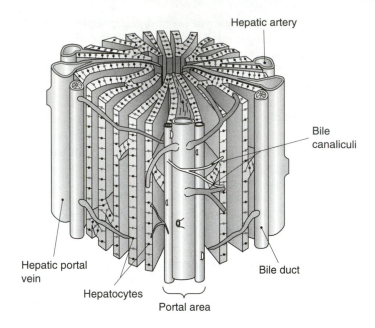

▶ **FIGURE 6–2**
View of the lobular organization of the liver.

the gut. These nutrients are acted on in certain ways that constitute the separate function of the liver (Figure 6–2). The bile produced leaves the liver through the *hepatic duct* and enters the duodenum through the common bile duct.

Without question, the liver, with its numerous metabolic and regulatory roles, is one of most important organs of the body. Without the liver we would die within 24 hours.

Functions

Although hundreds of different functions have been identified, the liver has seven principal functions:

1. *Produces bile:* The liver's main role as a digestive exocrine gland is the production of a yellow-green water solution containing bile salts, pigments (chiefly *bilirubin,* a breakdown product of hemoglobin), cholesterol, phospholipids, and a variety of electrolytes. Formed at the rate of about .95 liters (1 quart) a day, the bile salts and acids help to emulsify fat by activating enzymes that digest fat, by making fatty acids soluble to allow absorption of vitamins, and helping to exert antibacterial effects.
2. *Maintains glucose, amino acids, and fatty acid balances:* The liver regulates blood sugar levels by glycogen synthesis and is responsible for converting glucose to glycogen, and glycogen back into glucose. As discussed earlier, glucose is blood sugar that provides energy for all living things. Glucose is the major fuel used for providing ATP used by all cells of the body.
3. *Stores:* (a) glycogen and lipids, (b) vitamins A, D, E, and K, and (c) minerals such as iron, copper, potassium, cobalt, and zinc. Blood is also stored in the liver.
4. *Removes old RBCs from the bloodstream.*
5. *Produces plasma proteins and certain clotting factors.*

6. *A variety of metabolic functions:* Fatty acids are metabolized, ketone bodies are formed, useful cholesterol is formed, and *albumin* (which holds fluids in the bloodstream) is produced.

7. *Detoxifies certain hazardous substances:* In the liver, many toxic substances that could be potentially harmful to the body are converted into a water-soluble form. In this form they may be excreted into the bile (see Figure 2–11), destroyed by liver cells, conjugated, or even temporarily stored in the cells of the liver known as *Kupffer cells.* Through these activities the liver can alter a chemical's toxicity, its binding properties, its distribution throughout the body, and also its duration within the body.

   Most often, as discussed in Chapter 2, this biotransformation is beneficial to the body; however, in some instances it may lead to a more toxic substance or carcinogenic chemical as a by-product. *Necrosis,* or the death of cells, tissues, and organs, can result from the production of such toxic substances. The insecticides parathion and malathion are examples of how biotransformation in the liver may create or activate an even more toxic form of the chemical. The diseased liver cannot carry out its functions and, therefore, the body's threshold is lowered to almost every other toxic chemical.

Diseases of the Liver

Damage to the liver, typically, can result in many other types of injury. The exact impact depends primarily on the chemical involved, the dose and duration of exposure, and a host of other factors such as gender, diet, age, genetics, nutrition, and inhibitory chemicals. The effects on the liver can be either acute or chronic. Chemicals that target the liver are called *hepatotoxins.* The following is a partial list of diseases, the symptomatic effects, and some examples of chemicals that cause them:

1. *Lipid accumulation:* Once there has been an acute exposure, a number of agents cause abnormal increases of fats in the liver and other damages. Carbon tetrachloride, chloroform, ethanol, gasoline, phosphoric acid, tannic acid, tetracycline, and trichloroethylene are examples of agents producing fatty deposits within the liver.

2. *Choleostasis:* Inflammation of the biliary tree, with a coincidental decrease in bile flow and excretion, is known as choleostasis. This acute reaction is demonstrated by exposure to aminobenzylcaffeine, arsphenamine, chloropropomide, diazepam, erythromycin, imipramine, and mepazine.

3. *Necrosis:* Destruction or cell death can be the result of acute exposure to acetaminophen (nonaspirin), allyl alcohol, beryllium, galactosamine, tetrachloroethane, trichloroethylene, and urethane.

4. *Hepatitis:* Hepatitis is inflammation of the liver. Although commonly due to a virus, diffuse cell damage may be caused by acute exposure to acetohexamide, carbamazepine, ethionamide, halothane, imipramine, mercaptopurine, methyltestosterone, oxacillin, phenylbutazone, and trimethobenzamide.

5. *Cirrhosis:* This is a progressive disease most often due to chronic exposure. It is frequently characterized by widespread damage, is often associated with liver dysfunction, and frequently results in jaundice and portal hypertension. The malady of cirrhosis may be due to repeated minor insults, a single massive

injury, or by chronic doses of carbon tetrachloride, aflatoxin, or any of several other carcinogens. In both laboratory animals and in humans, the single most significant cause of cirrhosis is the ingestion of alcohol beverages or ethanol.

6. *Cancer:* Carcinogenesis is characterized by scattered nodules and tumors throughout the liver. Such malignancies are known to be elicited by PCBs, organochlorine pesticides, vinyl chloride, carbon tetrachloride, chloroform, and urethane.

Evaluation of Liver Injury

A number of tests can diagnose the state of the liver. Dye clearance tests, such as bromosulfophthalein and indocyanine green, are tests for clearance times. Another test of clearance times is the check for levels of bilirubin in the blood. Prothrombin clot time indicates the stability of the liver to produce blood-clotting proteins. Serum albumin tests demonstrate the ability of the liver to produce important proteins such as albumin. Tests that reveal the presence of certain enzymes in the blood, such as aminotransferase and serum alkaline phosphates, indicates that a liver cell's membrane is leaking, defective, and, therefore, potentially harmful to the liver.

Conclusion

The liver received its name because of the central role it plays in life itself. Because of its important role, nature has provided us with a surplus of liver tissue. We have more than we need, and even when we damage it or have part of it removed, it is one of the few body organs that can regenerate itself, and does so rapidly and easily.

Damage to the liver can result from a single event (an acute response), which may even result in rapid death. Chronic abuse of the liver may also result in liver damage and death, such as in regular intoxication from alcohol. Although the cells of the liver do not normally divide, tissue cells of the liver do retain their capacity to divide and repopulate the organ after their daughter cells have died. To make this effort worthwhile, continued exposure to and absorption of the hazardous material must be terminated.

THE KIDNEY

Each kidney is one of a pair of organs about 12.7 centimeters (5 inches) long and 2.5 centimeters (1 inch) thick. It is convex laterally and has an indentation near the middle that gives it the shape of a kidney bean. Another name for kidney is *renal.* The indentation is called the *hilus;* several structures, including the blood vessels feeding the kidney and numerous nerves enter the kidney at the hilus. At the top of each kidney is an *adrenal gland,* which is part of the endocrine system and is functionally separate from the kidney (See Figure 2–11, "Biliary Circulation.")

If you were to cut a kidney in half lengthwise, three distinct regions would become apparent: the *renal cortex,* which is in the outer region; the *renal medulla,* which is deep in the cortex; and the *renal pelvis,* which is a flat, basin-like cavity centered on the hilus.

The primary function of the kidney is to continuously cleanse the blood and adjust its composition. It is not surprising that the kidneys have a rich supply of blood with one-fourth of the body's blood supply passing through each minute. The artery that supplies the kidney with blood is the *renal artery,* which breaks down into smaller and smaller arteries called *interlobar arteries* that travel through the cortex. Blood then exits in a reverse direction by flowing in minute veins called *interlobar veins.* It continues to flow into larger and larger veins in the pelvis, and then finally returns to the heart via the *renal vein.*

The kidney has approximately one million miniature structures called *nephrons,* the kidney's structural and functional units that are responsible for the formation of *urine.* Each nephron consists of two structures: a knot of capillaries called the *glomerulus* and a *renal tube.* The closed end of the renal tube is enlarged and cup-shaped and completely surrounds the glomerulus; *Bowman's capsule* is the name given to this portion of the renal tube (Figure 6–3).

Kidney Function

The primary renal function is the elimination of body wastes, which is accomplished by excretion of urine. Urine is formed in a nonselective and passive process of filtration, whereby blood is forced through millions of nephrons where the formed elements in the blood are separated from the plasma. Nephrons control blood concentration and volume by removing water and solutes. They also regulate the pH balance between acids and bases within the body, and they remove toxic wastes from the blood during the process of filtration. Many nutrients and a few other specific substances are reabsorbed for essential body functions, and those not reabsorbed are eliminated from the body in the urine.

In addition to allowing needed substances to return to the blood and secreting wastes, the kidneys also produce enzymes to help regulate blood pressure (renin), and their hormone *erythropoietin* stimulates the production of red blood cells in the bone marrow.

Toxic Insults

The kidney has a high blood flow, and any chemical in systemic circulation will be delivered to the kidney in relatively high amounts. Response to such toxic insults may vary from nearly imperceptible aberrations to necrosis (cell death) and may be reflected in alteration of transport capability, renal failure, or complete suppression of the formation of urine. *Nephrotoxins* are chemicals capable of damaging the kidney and producing any of the full range of responses we just discussed. Depending on the magnitude of the insult, this damage can be reversible, permanent, or even lethal.

1. *Vasoconstriction:* Narrowing of the blood vessels, or vasoconstriction, can decrease renal blood flows and glomerular rates of filtration. This causes a reduction in urine flow and, thus, a buildup of blood urea nitrogen, which will lead to tissue destruction.
2. *Glomerular damage:* The glomerular element may be attacked directly by the nephrotoxin. This can result in a change of permeability, thus compromising the filtration process of the kidney.
3. *Tubular damage:* Tubular permeability may be altered by damage to the nephron cells lining the tubules. This affects the semipermeable membrane barrier of diffusion and, thus, suppresses the reabsorbative and secretory mechanisms.

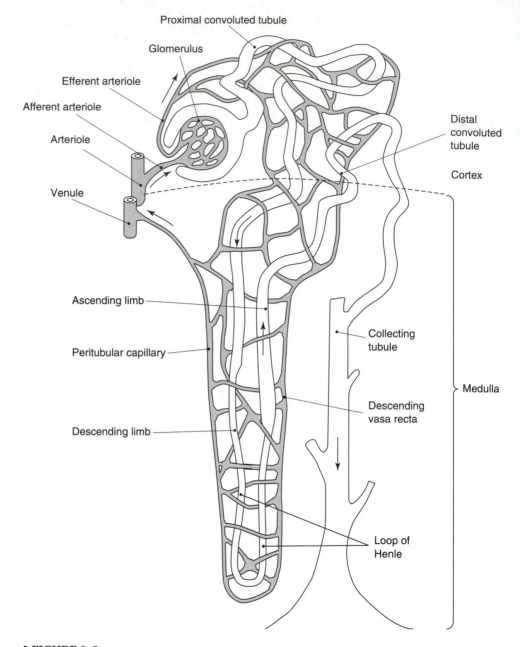

▶ **FIGURE 6–3**
A typical nephron and its blood supply.

4. *Acute renal failure:* Failure of the kidney from an acute dose is due to necrosis of the renal tubes. It can be caused by impaired blood flow and also by the effects of toxic chemicals. With proper diet and removal from exposure, the function usually returns after a variable period of time.

5. *Chronic failure:* Gradual deterioration of the kidney usually results from progressive renal disease, often continuing until complete dysfunction of the nephrons occurs. This type of dysfunction results in the derangement and imbalance of electrolytes, acid-base imbalance, retention of ions and other minerals normally excreted, retention of *urea* in the blood (nitrogen buildup and ammonia formation), and the retention of other assorted waste products.

Symptoms of Kidney Disease

Renal failure produces symptoms of weakness, loss of appetite, nausea, and vomiting. Weight gain often occurs because of fluid retention. Excess salt and water may also be retained due to lack of filtration by the failing kidney. Other conditions from poor filtration are decreased red blood cell production (due to reduced hormone production) that results in anemia, and blood pressure increases due to fluid retention and increased blood volume. Treatments include *hemodialysis,* to mechanically remove wastes from the blood; extreme cases of kidney failure can, hopefully, be corrected by *kidney transplant.*

Nephrotoxins

A variety of substances are nephrotoxic. Most heavy metals can injury the kidney, including cadmium, mercury, arsenic, bismuth, chromium, platinum, thallium, and uranium. Additionally, the halogenated hydrocarbons, especially carbon tetrachloride and chloroform, are nephrotoxins. Other chemicals that do not produce direct effects on the kidney but do alter the metabolism of the kidney include PCBs and TCDD (tetrachlorodibenzo-p-dioxin) which is a relatively harmless substance but is metabolically altered (biotransformed) within the kidney and produces nephrotoxicity.

Conclusion

The kidney is critical to the proper maintenance of blood because it is responsible for excreting wastes and toxins from the body. The kidney collaborates with many other organ systems to maintain homeostasis. A significant fraction of the energy our bodies expend is spent to protect the kidney and allow it to perform its functions. Although the kidney and the nephrons do not regenerate, the nephrons can grow and we can survive on as little as 10% of the nephrons found in one kidney.

THE EYE

Given the importance of vision, the eye is an organ of the body that is greatly underprotected. Perhaps this is not surprising when one considers the slowness of evolution in relation to the speed of the industrial revolution. The human eye was never designed for protection from the hazards of the modern workplace and environment. With comparatively few defense mechanisms, or regenerative processes, the eye is, undoubtedly, the most vulnerable of all the major body organs to occupational injury. WEAR YOUR SAFETY GLASSES!

Anatomy

Figure 6–4 illustrates the anatomy of the eyeball.

The eye is contained in a socket of bone. It is surrounded by a relatively thin layer of cushioning tissue. In the front, the eyelid closes over the eye. Although thin, the skin on the front of the eyelid offers the same type of protection to the environment as on other parts of the body. The eyelashes offer marginal, at best, protection for debris blown toward the eye. Reflexes are designed to close the eyelid before impact. The inner surface of the eyelid is lined with a moist mucous epithelium, the conjunctiva, which keeps the surface of the eye moist and clean. Tears, produced by eye ducts above and to the side of the eye, flow across and down. Besides providing moisture, tears contain nutrients and also contain *lysozyme,* an antibacterial enzyme, and antibodies to help defend the eye. Tears are actually an excess of lacrimal secretions caused when the eyes are irritated by foreign objects or chemicals, injury from trauma, and when we are emotionally upset. The tears cleanse and protect the eye surface in addition to providing moisture for lubrication.

The *eyeball* itself is a hollow sphere that is filled with fluids called *humors,* which help to maintain its shape. The eye is divided by the *lens,* which is the focusing apparatus of the eye, into two primary chambers: the aqueous chamber in front and the vitreous chamber behind the lens.

The eyeball is surrounded by three coats. The following three layers, the tunics of the eyeball, are tissues that surround the internal structures and fluids of the eye.

▶ **FIGURE 6–4**
The structure of the eyeball.

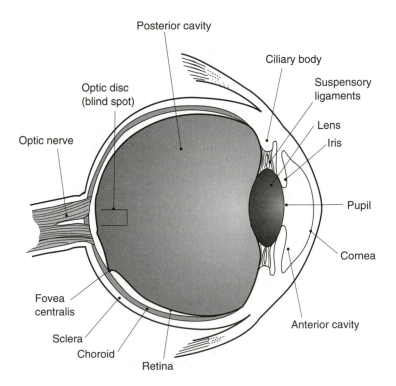

Sclera

The outermost tunic is a coat of dense fibrous connective tissue. Most of this is the "white" of the eye and is analogous to the tough membrane, the *dura,* which surrounds the rest of the CNS. In the front, this coat has been modified extensively to nonpigmented epithelial cells. This structure is known as the *cornea,* and must be transparent to allow light rays to enter the eye without distortion. The cornea also bends the light rays so that they correctly focus on the back of the eye. A number of medical procedures are currently available to "sculpt" the cornea and eliminate the need for glasses. Damage to the cornea can reverse this process and cause blurry vision.

Choroid

The second coat is known as the choroid. This middle layer is the *vascular tunic* and is, therefore, responsible for providing blood flow to the eye. It is heavily pigmented and appears black in the normal eye. Albinos, organisms without pigmentation, do not have a black pigmented choroid, and the eyes appear red. This coloration is due to the presence of hemoglobin in the blood supply. The pigmentation is critical in preventing light from bouncing around in the back of the eye. Individuals with pigmentation problems tend to be light sensitive.

Toward the front of the eye, the choroid is modified to form the muscular structures called the *ciliary body* and *iris.* Tendons from the muscles of the ciliary body attach around the circumference of the lens. Constriction of these muscles alters the shape of the lens and, therefore, the point at which light is focused at the back of the eye. This mechanism, called *accommodation,* is used to "focus" on objects at various distances from the eye.

The iris is the prominently colored body described when giving the color of a person's eyes. The coloration is genetically determined. The iris is comprised of two sets of smooth muscle. There is a circumferential set which, when contracted, constricts the iris, making the opening (the pupil) smaller and restricting the amount of light entering the eye. The second set is radial. When contracted, the radial muscles increase the size of the pupil and allow a greater amount of light to enter the eye. These muscles are under autonomic nervous system control and tend to respond reflexively to light levels. A number of substances, such as belladonna, or conditions such as concussions, can alter the normal response.

The lens is a biconvex mass of transparent tissue. The lens focuses light on the back of the eye where the retina is located. For any number of reasons, the transparency of the lens may be altered. Normally, this is due to a restructuring and ordering of the proteins that comprise the bulk of the lens. Such a condition is termed a *cataract.*

Retina

The innermost sensory tunic of the eye is called the retina. The *retina* is made up of nerve tissues that are comprised of five layers of neurons. At the deepest layer are millions of receptor cells called the *photoreceptors* (rods and cones), which are responsible for black/white and color vision, respectively. The photoreceptors communicate via nerve cells with the pigmented choroid, and problems with this juncture lead to detachment of the retina and a subsequent loss of vision. The rods are pigmented by a red-hued rhodopsin, partially composed of vitamin A. Lack of vitamin A can lead to a reduction or loss of night vision.

Images of the world are projected on the retina. As with a video camera, light levels that are too high or too low lead to bad pictures. Light levels are automatically adjusted by the iris. The image is processed by the other four layers of the retina. They are responsible for the functions of contrast and edge discrimination. This processed visual information is relayed to the back of the head to the visual cortex of the occipital lobe. Here the image is mapped on the brain and the result is vision. Stimulation of the visual cortex, such as a blow on the back of the head, activates these maps and cause "stars" to be seen.

Chambers

In front of the iris is the aqueous chamber. In back of the iris, in front of the lens, is the posterior chamber. These two chambers are connected through the pupil, which is actually an absence of structure. The aqueous chamber is filled with a watery fluid called the *aqueous humor*, which nourishes the epithelial cells associated with the front of the eye. This fluid normally drains from the chambers through small channels, the *canals of Schlemm*, located around the circumference of the cornea. Blockage of these canals leads to an increase in interocular pressure. This condition is called *glaucoma*. In back of the lens is the vitreous chamber filled with *vitreous humor*, a gel-like substance that helps give the eyeball its shape.

Chemical Hazards to the Eye

The primary hazards to the eye occur when there is accidental splashing of substances onto the open, unprotected surface of the eye. Exposure to mists, vapors, and gases produce varying degrees of problems. Eye irritation ranges from minor inflammation to complete loss of vision. In some cases, a chemical may produce no damage to the eye itself, but enough may be absorbed to cause poisoning elsewhere in the body or to interrupt the visual process.

Irritant chemicals can give rise to *acute keratitis*, which begins as pinpoint holes in the cornea. The cornea may quickly break down into larger erosions. Some industrial chemicals (lacrimals) can cause excessive watering of the eyes, or *lacrimation*, due to irritation of the mucous membranes. Other effects include disturbances of vision, discoloration, paralysis of the eye muscles, double vision, optic atrophy, and either temporary or permanent blindness.

Burns of the eye are quite common because of exposure to caustic solutions. Caustics tend to be even more damaging than acids to the eye and skin because acids dilute in water and are readily washed away. Bases, on the other hand, rapidly react with cornified tissue and adhere tenaciously. They continue to soak into the tissue as long as they remain in contact. If exposed, flush the surface of the eye at least 15 minutes!

Agents That Damage the Eye

The eyes are affected by the same chemicals that damage the rest of the skin. However, the skin of the eye is much thinner and more sensitive than skin elsewhere on the body. Additionally, the noncornifying transparent cornea lacks much of the rest of the skin's protective mechanisms. The following types of materials can damage the eye by direct contact.

Acids

The pH and the protein-combining capacity of acids determine the likeliness of damage to the eye. The burns initiated from contact during the first few moments indicate the long-term damage to be expected from the exposure. The most common industrial acids are sulfuric, picric, tannic, and hydrochloric. Immediately flush your eyes with copious amounts of water if exposed to these acids!

Alkalis

Initially perceived as causing slight damage, exposure to caustics and bases leads to delayed effects. Delayed effects include ulceration, perforation of the cornea, and clouding of the lens. The pH and duration of exposure have more bearing on the amount of damage than the type of alkali. Some common alkalis are sodium hydroxide (caustic soda), potassium hydroxide, ammonium hydroxide, calcium oxide, and lime. If exposed to these caustics, flush with water for at least 15 minutes!

Organic Solvents

Organic industrial solvents include alcohols, benzene, toluene, hexanes, and acetone. Other common solvents are gasoline, paint and varnish thinners, and plastic pipe cements. These organic compounds dissolve fats, cause pain, and dull the cornea. The extent of damage depends on the concentration, duration of exposure, and temperature of the solvent.

Miscellaneous Substances

A variety of sources for other substances may affect the eye. Heavy metals, for example, dusts from lead, copper, and tin, as well as ions that come with protein functional groups, for example, silver proteinate and mercury oxide, are toxic. Dusts of hydroquinone, dichloroethane, and phenylenediamine (from eye makeup) are toxic. Quinacrine and chlorpromazine are toxic systemic drugs. Dinitrophenol, an antiobesity drug, can cause cataracts. Thallium, used as an insecticide, and naphthalene vapors negatively affect the eye.

Conclusion

Vision is probably the most important sense; 70% of all sensory receptors in the body are in the eye. Protect your eyes at all times with the use of proper personal protective equipment, which is mandatory when working with hazardous substances. No matter how remote the possibility of exposure may seem, an ounce of prevention is worth a pound of cure. BE PREPARED!

THE IMMUNE SYSTEM

The *immune system* refers to a number of mechanisms that the body uses to defend itself from foreign (exogenous) substances, and the *lymphatic system* constitutes a major portion of these mechanisms for defense. Defense mechanisms also include a variety of white blood cells (WBCs), especially the leukocytes and the macrophages. In this section we will briefly discuss WBCs and their contribution, review the operation and function of the lymphatic system, and discuss the remaining important aspects of the body's immune response.

White Blood Cells (WBCs)

Leukocytes are white blood cells that form a protective and "movable" army that helps the body defend itself against damage from bacteria, parasites, viruses, tumor cells, and hazardous substances. While red blood corpuscles are confined to the bloodstream, WBCs are able to slip into and out of the blood vessels in a process called *diapedesis;* that is, they use the circulatory system to travel to areas of the body where they are needed for either inflammatory or immune responses. These WBCs are able to locate these areas of need by responding to certain chemicals that diffuse from the insulted or damaged cells. This capability is called *positive chemotaxis.* Whenever WBCs are called into action, the body speeds up its production. *Leukocytosis* is a term for high counts of WBCs, generally indicating a bacterial or viral infection somewhere in the body, and is a normal response. However, a cancer of bone marrow results in an increase in white blood cells, a diseased state called *leukemia. Leukopenia* is an abnormally low count of WBCs usually caused by certain drugs, such as corticosteroids. *Lymphocytes* are a special type of WBC that are strategically located within the lymph nodes and respond to foreign substances in the lymphatic system.

The Lymphatic System

As discussed briefly before, the plasma of the blood circulates from the heart through arteries to capillary beds. In the capillary beds, fluids are able to cross the capillary wall and permeate the interstitial spaces around the cell tissues, where it is called *interstitial fluid.* Interstitial fluid diffuses into small single-ended (blind) tubes, the lymphatic capillaries, where it is now called lymph. The blind capillaries coalesce into larger tubes. Large tubes join at enlargements, nodules (avascular), and nodes (vascular). This fluid is filtered within the lymph. The lymphatic system also includes the system of tubes that circulate this fluid back to the veins of the circulatory system. All along this system of tubes are the many *lymph nodes* that help protect the body by filtering out bacteria and other substances. The lymph system also includes the spleen, which filters the blood, and the thymus glands. These glands are important in the development of our immune response and the development of immunity (mentioned in the following section).

The lymphatic tissue contains numerous *lymphocytes,* a special type of leukocyte or formed element of the blood, within each lymph node or nodule. When the lymph tissue has filtered the fluid and trapped a number of foreign substances, the lymph glands become swollen and tender: most of us have had swollen glands during an active infection. The lymphocytes themselves are capable of actually destroying exogenous microorganisms and substances. Also within the lymphoid organs are *macrophages,* which literally means "big eaters." Their function is to engulf foreign particles and, when activated by T cells, to become insatiable phagocytes.

The Immune Response

The immune system is also a functional system, as opposed to an organ system, that recognizes foreign molecules *(antigens)* and acts to inactivate or destroy them. Factors in the blood that protect it from future attacks of antigens are known as *antibodies.* Thus, immunity is the body's ability to resist damage or disease from exogenous materials and is categorized as either nonspecific immunity or specific immunity. When there is an insult or exposure to foreign or harmful substances, a set of events occur within the body to defend itself and the same set of events occur regardless of the nature of the insult. This set of events is called *nonspecific immunity.* With *specific immunity,* the body learns after the first exposure; the response to the second exposure of the same substance is faster and stronger than the response to the first exposure because the body has prepared itself and is ready for the insult. For example, a vaccination prepares your specific immune system for a particular invading organism by injecting the essence of it into the body. The body responds with a specific set of events to try to identify the exogenous material by creating an antibody. The memory of that response allows the body to get ready for the second invasion. When the material invades again, the response is so rapid and effective that the material is usually destroyed even before any symptoms develop. This immune system response is possible because of *specificity* (it recognizes the essence and acts against particular pathogens or foreign substances) and *memory* (it recognizes the substance, recalls the antibody from the library, and mounts an even stronger defense against the previously encountered pathogens).

Nonspecific Immune Response

Nonspecific response occurs after trauma, burns, bacterical invasions, and chemicals. It consists of corralling off the area and increasing blood flow to the area, which results in redness and swelling—*inflammation.* Frequently, a side effect is an increase in body temperature, or *fever.* WBCs, basophils, which migrate to the area, release histamines, which cause inflammation and fever, and substance "P," a protein that causes pain. These symptoms are typically countered by taking antihistamines and anti-inflammatories such as aspirin.

Related Immune Response

The immune system guards against foreign substances by blocking their entry or by preventing their distribution throughout the body. This may be accomplished by one of the following mechanisms.

Mechanical Barriers   These barriers prevent entry of microorganisms into the body or physically remove them from the surface. Examples of mechanical mechanisms include both surface membranes (such as the skin) and mucous membranes that line all the body cavities that are open to the exterior, including the digestive, respiratory, urinary, and reproductive tracts. These barriers are made up of epithelium tissues that also produce some fluids and protective chemicals that kill pathogens, help repair tissue, and fight off other exogenous materials. For example, substances are washed from the eyes by tears, from the mouth by saliva, and from the urinary tract by urine. In the respiratory tract mucous membranes lined with miniature cilia help remove substances trapped by the mucus and carried by the mucociliary escalator to the back of the throat where they are swallowed or spat out. Coughing and sneezing also remove microorganisms from the respiratory tract.

Chemicals Various chemicals are used by the body to act directly with or against other foreign substances. Lysozymes (enzymes), sebum (oil), and mucus (glue) are found on the surface of cells and kill or prevent entry of foreign organisms. *Histamine* is released from white blood cells causing vasodilation or enlargement of the blood veins, increases in vascular permeability, stimulation of glandular secretions, contraction of smooth muscles, and the attraction of white blood cells to the area. *Kinins* are polypeptides derived from plasma proteins that also cause vasodilation, stimulate pain receptors, and attract WBCs to help fight acute infection. *Interferon* is a protein that interferes with virus production and infection. *Prostaglandins* and *leukotrienes* are lipids that cause smooth muscle relaxation, vasodilation, and attract other cells to the area. *Pyrogens* are chemicals that stimulate fever production. Foreign chemicals have the potential to stimulate, emulate, or depress the body's natural chemical activities.

Phagocytosis Phagocytosis and immune system response can involve the participation of a *complement system* of 11 proteins that comprise approximately 10% of the globulin portion of the serum. When activated, this complement system promotes the development of phagocytes and attracts them.

Inflammation Inflammation mobilizes the immune system and isolates microorganisms until they can be destroyed. Natural killer cells (NK) are lymphocytes produced in red bone marrow that target general classes of cells, such as tumor cells, rather than any specific cell. This is a nonspecific response; NK do not exhibit a memory, and they use a variety of methods to kill their target cells, including the release of chemicals that damage cell membranes (Table 6–1).

Specific Response Function

The specific response portion of the immune system learns how to recognize, respond to, and destroy intruders/invaders; it also remembers the particular substance or material (antigen). *Haptens* are small molecules that, when combined with larger molecules, stimulate a specific immune response. B cells are associated with antibody-mediated immunity. Through a complex series of events, the essence of the insult is recognized and an antibody is prepared. A library of antibodies is maintained to allow a rapid response after future invasions.

Immune Response-Altering Agents

Industrial workers and consumers are exposed to many materials capable of producing hypersensitivity in the organs of the body. *Sensitization,* simply put, is a response by the immune system from exposure to a particular type of antigen known as an *allergen.* Exposure to an allergen causes the body to manufacture antibodies that will react with the allergen. Once antibodies are formed, the person is said to be sensitized and any future exposure to the allergen (even in much smaller amounts)

❱ TABLE 6–1
Summary of nonspecific body defenses.

Category and Associated Elements	Protective Mechanism
Surface membrane barriers	
Intact skin (epidermis)	Forms mechanical barrier to prevent entry of pathogens and other harmful substances into body.
Acid mantle	Skin secretions make epidermal surface acidic, which inhibits bacterial growth.
Keratin	Provides resistance against acids, alkalis, and bacterial enzymes.
Intact mucous membranes	Form mechanical barrier to prevent entry of pathogens.
Mucus	Traps microorganisms in respiratory and digestive tracts.
Nasal hairs	Filter and trap microorganisms in nasal passages.
Cilia	Propel mucus carrying debris away from lower respiratory passages.
Gastric juice	Contains concentrated hydrochloric acid and protein-digesting enzymes that destroy pathogens in stomach.
Acid mantle of vagina	Inhibits growth of bacteria and fungi in female reproductive tract.
Lacrimal secretion (tears); saliva	Continuously lubricate and cleanse eyes (tears) and oral cavity (saliva); contain lysozyme, an enzyme that destroys microorganismas.
Cellular and Chemical Defenses	
Phagocytes	Engulf and destroy pathogens that breach surface membrane barriers; macrophages also contribute to immune response.
Natural killer cells	Promote cell lysis by direct cell attack against virus-infected or cancerous body cells; do not depend on specific antigen recognition.
Inflammatory response	Prevents spread of injurious agents to adjacent tissues, disposes of pathogens and dead tissue cells, and promotes tissue repair; chemical mediators released attract phagocytes (and immuno-competent cells) to the area.
Antimicrobial chemicals	
Interferons	Proteins released by virus-infected cells that protect uninfected tissue cells from viral takeover; mobilize immune system.
Complement	Lyses microorganisms, enhances phagocytosis by opsonization, and intensifies inflammatory response.

results in an allergic reaction such as hives, dermatitis, runny nose, headache, or asthma. For some people this sensitization may be just an annoyance, and for others it may be debilitating but fortunately deaths from exaggerated allergenic reactions (anaphylactic shock) are rare. Certain substances cause particular organs to become sensitized, and a number of them are described in the following sections.

Lung Disease

Asthma is one of the major types of hypersensitivity observed in industrial settings. Responsible chemicals include toluene, bacillus subtilus (a bacterial detergent enzyme preparation), and numerous other industrial compounds.

Contact Dermatitis

Allergenic contact dermatitis is evoked by a variety of substances including poison ivy, drugs, cosmetics, certain metals (nickel and chromates), formaldehyde, disinfectants, inks, wood products, resins, and plasticizers such as trimellitic anhydride. Other chemicals include platinum, mercury, beryllium, sorbic acid, paraben esters, and ammonia compounds.

Autoimmunity

Autoimmunity is the failure of the body to recognize its own antigens, and therefore it responds as it would to any foreign substance. Severe cases are lethal; mild cases lead to hypersensitivity to some foreign substances. The leading substances associated with this disorder include penicillin, methyldopa, salicylates such as acetylsalicylic acid (aspirin), quinidine, and tetracycline.

Immunosuppressive Chemicals

Immune suppressive chemicals include benzene, halogenated aromatic hydrocarbons such as polychlorinated biphenyls (PCBs), dibenzodioxins such as tetrachlorodibenzo-p-dioxin (TCDD), hydrocarbon combustion by-products, urethane (ethyl carbamate), phenol diester, organophosphate insecticides, carbamate insecticides, organo-chlorine insecticides such as DDT, airborne pollutants and metals, and heavy metals such as lead, cadmium, mercury, arsenic, and nickel.

Immunodeficiencies

Currently, the most important and most devastating immunodeficiency is human immunodeficiency virus (HIV), which in recent years has demanded an unknown number of research hours and study. The HIV actually attacks the immune system and specifically destroys helper T cells. Eventually the whole immune system is turned topsy-turvy. When the organism itself is finally overwhelmed by the HIV, the victim is said to have Acquired Immune Deficiency Syndrome, or AIDS.

Conclusion

As amazing as the immune system is, it has one major shortcoming: To activate the specific immune response, the system must first "meet" each foreign substance before it can protect the body against it. A second shortcoming is its own susceptibility to many substances that are common in occupational settings. Awareness of and protection from these hazards are essential to maintaining a safe and healthy work environment. Table 6–2 provides a summary of the functions of the cells and molecules involved in immunity.

▶ **TABLE 6–2**
The functions of the cells and molecules involved in immunity.

Element	Function in the Immune Response
Cells	
B cell	Lymphocyte that resides in the lymph nodes, spleen, or other lymphoid tissues, where it is induced to replicate by antigen-binding and helper T-cell interactions; its progeny (clone members) form memory cells or plasma cells.
Plasma cell	Antibody-producing "machine"; produces huge numbers of the same antibody (immunoglobulin); represents further specialization of B-cell clone descendants.
Helper T cell	A *regulatory* T that binds with a specific antigen presented by a macrophage; upon circulating into the spleen and lymph nodes, it stimulates the production of other cells (killer T cells and B cells) to help fight the invader; acts both directly and indirectly by releasing lymphokines.
Killer T cell	Also called a cytotoxic T cell; recruited by antigen presented by a macrophage; activity enhanced by helper T cells; its specialty is killing virus-invaded body cells, as well as body cells that have become cancerous; involved in graft rejection.
Suppressor T cell	Slows or stops the activity of B and T cells once the infection (or attack by foreign cells) has been conquered.
Memory cell	Descendant of an activated B cell or T cell; generated during the initial immune response (primary response); may exist in the body for years thereafter, enabling it to respond quickly and efficiently to subsequent infections or meetings with the same antigen.
Macrophage	Engulfs and digests antigens that it encounters, and presents parts of them on its plasma membrane for recognition by T cells bearing receptors for the same antigen; this function, antigen presentation, is essential for normal, cell-mediated responses; also releases monokines chemicals that activate T cells.
Molecules	
Antibody (immunoglobulin)	Protein produced by a B cell or its plasma cell offspring, and released into the body fluids (blood, lymph, saliva, mucus, etc.), where it attaches to antigens, causing neutralization, precipitation, or agglutination, which "marks" the antigens for destruction by phagocytes or complement.
Lymphokines	Chemicals released by sensitized T cells: • Migration inhibitory factor (MIF)—inhibits migration of macrophages, thus keeping them in immediate area. • Macrophage activating factor(MAF)—"activates" macrophages to become killers. • Interleukin II—stimulates T cells to proliferate. • Helper factors—enhance antibody formation by plasma cells.

(continued on page 138)

◗ **TABLE 6–2**
(continued)

	• Suppressor factors—suppress antibody formation or T-cell-mediated immune responses.
	• Chemotactic factors—attract leukocytes (neurophils, eosinophils, and basophils) into inflamed area.
	• Lymphotoxin (LT)—a cell toxin; causes cell lysis.
	• Gamma interferon–helps make tissue cells resistant to viral infection; also released by macrophages.
Monokines	Chemicals released by activated macrophages:
	• Interleukin I—stimulates T cells to proliferate and causes fever.
	• Interferon—helps virus-infected tissue cells by preventing replication of virus particles that have invaded them.
Complement	Group of bloodborne proteins activated after binding to antibody-covered antigens; when activated, complement causes lysis of microorganism and enhances inflammatory response.

SUMMARY

This chapter continued the study of systemic toxicology and included discussions of the following target organs and target organ systems:

◗ The respiratory system (a) delivers oxygen to the body's tissues for use in numerous metabolic processes and (b) removes carbon dioxide and other waste products from those processes. Respiratory toxins are substances that cause respiratory disorders and interrupt these primary functions. Entry of hazardous substances by route of the respiratory system is the most common form of contamination and occurs primarily in occupational settings.

◗ The liver is an essential organ in detoxifying and biotransforming hazardous materials. The liver is also a target of hazardous substances called hepatotoxins.

◗ The primary function of the kidney is elimination of body wastes by producing and excreting urine. The kidney is essentially a filter and, because large volumes of blood pass through it every minute, it is also a prime target for hazardous substances. Toxic insults of the kidney are called nephrotoxins and can cause a variety of kidney malfunctions and failure.

◗ The eye is an organ of the body that is greatly underprotected and vulnerable to contamination from hazardous substances. Agents that damage the eye include corrosive acids and bases, solvents, and a wide variety of substances that can be dissolved in the membranes of the eyes, which are richly vascularized, providing easy access to the central nervous system.

◗ The immune system is the final system discussed in this chapter. This system protects our body from the invasion of foreign/harmful substances by a variety of means. There are mechanical barriers, chemical interactions, and phagocytic responses that are nonspecific to any type of foreign insult. Specific responses are those that occur when the immune system recognizes, responds to, destroys, and remembers particular substances or materials to be fully ready for a second insult. Substances that are harmful to this system are called immunotoxins.

STUDY QUESTIONS

1. What is the purpose of the respiratory system? Name each component part.
2. What is the purpose and the function of the liver? Name three chemicals or chemical groups that adversely affect this organ.
3. What is the purpose and the function of the kidney? Name three chemicals or chemical groups that adversely affect this organ.
4. What is the purpose and the function of the eye? Name three chemicals or chemical groups that adversely affect this organ.
5. What is the purpose and the function of the immune system? Name three chemicals or chemical groups that adversely affect this system.

7

Toxic Agents

COMPETENCY STATEMENTS

This chapter discusses the various toxic agents that have harmful effects on the body. After reading and studying this chapter, students should be able to meet the following objectives:

▶ Identify, describe, and provide several examples of each of the following groups and subgroups of chemicals: (a) pesticides, including insecticides, herbicides, fungicides, and rodenticides; (b) fumigants; (c) metals; (d) solvents and their vapors; (e) animal toxins; and (f) plant toxins.

▶ Briefly describe the organs or organ systems targeted by each of the chemicals listed above.

INTRODUCTION

The effect of synthetic and natural substances on human health is a common concern in modern life. It has been suggested that many diseases are the result of environmentally encountered materials. Constant exposure to a low concentration of a substance, or several substances, creates a "chemical stress" on the body's defenses. While the levels of the substances are below regulated thresholds, they may be high enough to affect sensitive or sensitized individuals. Additionally, the chemical stress may be just enough that, in combination with other factors, a person's level of wellness is decreased. At some point a lack of wellness will become an illness. This chapter provides an abbreviated discussion of pesticides, metals, solvents, and vapors and the effects of animal and plant toxins.

PESTICIDES

Pesticides are unique because they are added deliberately to the environment for the purpose of killing some forms of life. Ideally, they are prepared to target a specific pest (insect, rodent, or other) but, instead, are generally toxic to many nontarget species, including humans. Acute poisonings are usually caused by occupational exposures, carelessness, or misuse. Therefore, it follows that the principal routes of exposure are by improper care or use of personal protective equipment (PPE). Proper protective equipment should always be worn by workers. Other frequent dosage mechanisms are orally, by accidental ingestion, and as suicide attempts. Numerous incidents of poisoning have resulted from eating food that has become contaminated with pesticides during shipping or storage.

Insecticides

Various classes of insecticides will be considered here. The larger classification of pesticides is sometimes used interchangeably with insecticides. However, there are numerous classes of pesticides; insecticides are just one class.

Organophosphorus Insecticides

Organophosphorus insecticides are frequently involved in serious poisonings of human beings. The first organophosphates were developed in Germany during World War II as the original nerve gas chemical warfare agents. Dimethyl phosphonoamidocyanidate (tabun), isopropyl methylphosphonofluoridate (sarin), and, an analogous chemical form, tetraethyl pyrophosphate (TEPP) were among these. Sarin was the first to be used as an insecticide. However, because it was far too toxic to be used around mammals, parathion was developed and has the distinction of being responsible for most of the fatal poisonings by a pesticide.

The immediate cause of death from an acute exposure is respiratory failure. Organophosphates also inhibit the function of acetylcholinesterase, which allows acetylcholine to accumulate in the nervous system and related organs and causes the symptoms of tension, anxiety, restlessness, insomnia, headache, emotional instability, and neurosis. In more advanced stages, the poisonings cause slurred speech, tremor, general weakness, ataxia, convulsions, and coma. Three of the nerve gases—sarin, soman, and tabun—are still important military weapons. These substances are liquids at room temperature, tasteless, and colorless (tabun can be brownish). If ingested, they kill instantaneously. They are absorbed quickly from the lungs after inhalation. One drop of sarin on the skin is lethal to humans in only 15 minutes.

Other chemicals of this class, with similar symptomatology in toxic doses, are mevinphos, disulfoton, chlorfenvinphos, dichlorvos, diazinon, dimethoate, trichlorfon, and chlorothion. Organophosphates are rapidly biotransformed and will degrade in a relatively short time. Because they tend to be excreted quickly, chronic poisoning by accumulation of the compounds usually does not occur. However, with even a small amount by skin contact, inhalation, or ingestion and absorption, death can occur with little or no warning.

Carbamate Insecticides

Carbamate insecticides are similar to the organophosphates in action. They act on the acetylcholinesterase but, unlike the organophosphates, the carbamates have very low skin (dermal) toxicities. The signs and symptoms of poisoning are tearing in the eyes (lacrimation), salivation, contraction of the pupil (miosis), convulsion, and death. Carbaryl, one of the least toxic carbamate insecticides, has been demonstrated to have teratogenic (genetic birth defects) consequences in laboratory animals. Other examples of the carbamates are Baygon, Propoxur, Mobam, Temik, Aldicarb, and Zechtran.

Organochlorine Insecticides

Organochlorine insecticides are a group that includes the chlorinated ethane derivatives of which DDT is the most well known. The cyclodeines, which include chlordane, aldrin, dieldrin, epatachlor, and toxaphene and the hexachlorocyclohexanes, such as lindane, are also of this class.

The organochlorines have come into disfavor because of their persistence in the environment. They also tend to accumulate in biologic and nonbiologic media. As a class, they are less toxic than both the organophosphorus insecticides and the carbamate insecticides. However, because of this bioaccumulation property, they have the potential to be many times more toxic after extended periods of use. Classed as neuropoisons, their mechanism of action does not affect cholinesterase or mimic acetylcholine. Their specific molecular mechanism of action is currently under investigation.

Botanical Insecticides

Botanical insecticides are not necessarily any less toxic to mammals than synthetic insecticides are. Those made from natural derivatives include nicotine, which interferes with pathways of the central nervous system. Rotenoids, including pyrethrum, are common household insecticides. They are valuable because of their rapid "knock-down" action, or ability to rapidly paralyze and exterminate household pests. A number of the synthetic pyrethroids are extremely toxic to aquatic animals.

Herbicides

Chemicals that destroy weeds are designed for markedly different physiological targets in plants. Nonetheless, these have accounted for many fatal poisonings of man.

Chlorophenoxy

Compounds such as dichloro (2,4–D) and trichlorophenoxyacetic acid (2,4,5–T), either as salts or esters, are the most familiar chemicals used as herbicides (to control broad-leafed and woody plants). At lower doses, to mammals, these chemicals cause stiffness of the extremities, paralysis, ataxia, and coma. Massive doses cause death by ventricular fibrillation (heart attack). Skin contact leads to chloracne in humans and is suspected of being teratogenic.

The combination of 2,4,5–T and 2,4–D (dichlorophenoxyacetic acid) is the highly publicized Agent Orange. Unfortunately, Agent Orange was usually contaminated with dioxins, which are very toxic. There is controversy over which effects are due to which ingredients. However, this mixture has been observed to produce liver damage, a disturbance in metabolism (porphyria), teratogenic effects, reproduction effects, immune suppression, tissue necrosis, and increased tumor formation (neoplasms). Experiments with these products are still underway to delineate and characterize the full spectrum of their adverse biological actions.

Dinitrophenol

These compounds cause an increase in body temperature. Related symptoms reflect an increased metabolism which, if unchecked, can result in fatal hyperthermia. Chronic exposure to dinitroorthocresol may lead to increased metabolic rates, permanently stain the conjunctiva of the eye yellow, and lead to cataract development. Dinitrobutylphenol may produce symptoms similar to those of dinitrophenol.

Biparietal Compounds

The best-known compound of this class is paraquat. Paraquat is responsible for a high number of fatalities, either accidental or suicidal. Diquat is similar in structure to paraquat; however, it targets the gastrointestinal tract, liver, and kidney whereas paraquat principally damages the lungs.

Carbamate Herbicides

This group includes propham and barban, which have relatively low acute toxicities, but still need to be treated with caution.

Fungicides

Fungicides are a class of compounds that control or kill fungi, yeast, and their spores. The group of mercury-containing fungicides is of great potential hazard to humans. The mercurials have been responsible for numerous deaths and permanent neurologic damage. In the 1950s and 1960s in Japan, industrial releases of methyl and other mercury compounds into Minamata Bay were followed by accumulation of the mercury in edible fish. This contamination caused a major epidemic of methylmercury poisoning. The largest recorded epidemic of methylmercury poisoning took place in Iraq in 1971 and 1972; over 6,000 patients were hospitalized and there were at least 500 deaths. The source of methylmercury was bread containing wheat imported as seed grain that had been treated with methylmercury fungicide.

Other fungicides are suspected of being teratogens because of chemical structures similar to thalidomide (a known birth-defect-causing agent). In laboratory experiments, Captan has been shown to be teratogenic (in rabbits, rats, and hamsters), as well as mutagenic and carcinogenic. Substituted aromatics, such as pentachlorophenol (PCP) and pentachloronitrobenzene (PCNB), are used in great quantities as fungicides (over 22.7 million kilograms, or 50 million pounds per year). These chemicals must be used with extreme caution. Improper use has resulted in numerous reported cases of fatal poisonings. Dithiocarbamate fungicides have a low acute toxicity with little evidence of human injuries from exposure. However, laboratory reports indicate possible teratogenicity and carcinogenicity. Nitrogen heterocyclic fungicides, such as benomyl and thiabendazole, have caused acute dermatitis in greenhouse workers.

Rodenticides

The major hazard posed by rodenticides is acute toxicity either by accident or by suicide. Some rodenticides are:

1. *Warfarin:* Warfarin, 3-alpha-acetonylbenzyll-4-hydroxycoumarin, is the most widely used rodenticide. It is a good rat poison because it requires repeated dosing for toxic effects to develop. Therefore, it requires repeated dosages to poison humans. Warfarin works as an anticoagulant, inhibiting the synthesis of prothrombin, with the subject bleeding to death (hemorrhage).
2. *Red squill:* Red squill is a plant glycoside that, in large doses, causes the heart ventricles to beat irregularly.
3. *Norbormide:* Norbormide causes smooth muscles to contract and shut down blood flow.
4. *Sodium fluoroacetate and fluoroacetamide:* These compounds are very toxic and inhibit cellular metabolism (Krebs cycle blockage).
5. *Strychnine:* Strychnine is an alkaloid of the nux vomica plant; it is a potent convulsant.
6. *Zinc phosphide:* Zinc phosphide causes gastrointestinal irritation.

FUMIGANTS

Fumigants are pesticides typically of a gaseous form. Therefore, they will penetrate areas that are otherwise inaccessible for pesticide application. Because of their gaseous nature, they particularly are potential inhalation hazards. The fumigants used to protect foodstuffs are of the greatest human hazard potential. Fumigants include acrylonitrile, carbon disulfide, carbon tetrachloride, chloropicrin, ethylene dibromide, ethylene oxide, hydrogen cyanide, methyl bromide, and phosphine.

METALS

Humans are composed of the elements of the earth, and some of these are small quantities of metals included in the group known as *trace elements*. These elements are needed by the body but only in very minute quantities and thus are also called *micronutrients*. Examples of micronutrients include zinc, iron, iodine, selenium, boron, copper, molybdenum, nickel, silicon, and vanadium.

All micronutrients are toxic to the body whenever the body is exposed to amounts greater than are needed (Remember: "The dose makes the poison"), and this is especially true of the heavy metals. The usual quantities in the body are less than 1 milligram per kilogram of body weight. Therefore, a number of metals that are plentiful in the environment constitute a potential threat to health as these substances are transported in the air, water, soil, and food. Humans may be exposed to these metals when rainwater, especially acid rain, dissolves rocks and carries the metallic component away in sediment. Plants and animals incorporate and concentrate metals into their food cycles; humans, in turn, eat the plants and animals and may also breathe dusts carrying small amounts of heavy metals.

Metals are probably the oldest toxins known to humans (Hippocrates [320 B.C.] described these), yet those of greatest concern today have only recently become identified and quantified in terms of toxicity. Lead has been known to be toxic since ancient times. It is frequently credited for bringing down the Roman Empire. Yet the use of lead in our society is ubiquitous; from gasoline to paint. Fortunately, these uses are being phased out; however, reservoirs from past use exist, and precautions must be taken.

Over the past 20 years our federal government has made great strides at eliminating lead from the environment. As the use of lead has decreased, so have the "safe" levels recommended by the regulators. Yet, in spite of all the progress that has been made, lead is still commonly used in gasoline in many states and other countries, from which it continues to pollute our environment. Eighty (80) of the 105 elements have been reported to be toxic metals (elements 106 to 109 are also probably toxic but they are short-lived laboratory curiosities). See Table 7–1.

Typically, the dose, or an estimate of absorption, of a metal is a function of both exposure time and concentration. Estimates of absorbed dose are usually verified from blood samples, urine, and hair (indicator tissues).

Factors Influencing Toxicity of Metals

The toxicity of metals is influenced by susceptibility factors of certain tissues, such as gastrointestinal, liver, and renal tubes, to metallic ions. The protective mechanisms of the cell, involved in detoxification or biotransformation of metals, are greatly influenced by the specific organs involved. Exogenous factors (diet, age, gender, and body-fat percentage) and lifestyle factors such as smoking and drinking influence the body's uptake and elimination of toxic metals. The chemical form of the metal ions and solubility of the metal in blood also influence toxicity.

Chelation

Chelation agents are those that can be introduced into the body to bind metals. These agents have an affinity for binding with metals and, therefore, for enhancing the body's ability to mobilize excretory mechanisms for the metals. Two examples of chelation agents are BAL, which chelates arsenic, lead, and mercury, and DMPS, (dimercapto-l-propanesulfonic acid), which chelates methylmercury, cadmium, copper, and nickel. Chelation agents are very toxic in their own right, as they remove essential metallic coenzymes and cofactors. Therefore, any chelation therapy requires strict monitoring by a physician.

Metals With Multiple Effects

Arsenic

Arsenic is widely distributed in nature and has a complex chemistry. The most prevalent forms are arsenic trioxide, sodium arsenite, arsenic trichloride, arsenic acid, and calcium arsenate. Arsenials are transported by industrial exhausts from the manufacture of, for example, pesticides, in the air and by water. Arsenic concentrates in hair and nails and causes harm to the lungs after inhalation. Ingestion may be fatal and symptoms include fever, anorexia, liver enlargement (hepatomegaly), upper respiratory tract symptoms, peripheral neuropathy, and gastrointestinal, cardiovascular, and hematopoietic effects. Arsenic is highly carcinogenic to the skin and in the lung; it is also teratogenic and mutagenic. Arsine gas is formed by reaction of arsenic with hydrogen; it is a by-product of refining nonferrous metals and is used in the semiconductor industry. Arsine gas causes renal failure and anemia, which are usually fatal.

▶ **TABLE 7–1**
Toxic metals.

Metal	Symbol	Uses	Toxic Forms	Toxic Exposure
Aluminum	Al	Chemical manufacturing, building materials, appliances, soldering, and many other uses	Powdered aluminum, aluminum oxide fumes from aluminum Aluminum alkyls (ignite spontaneously) and produce aluminum oxide smoke. Skin contact with aluminum alkyls causes burns. In industry one of the most important is triethylaluminum. Napalm is an aluminum soap with naphthenic acid and fatty acids.	Fibrosis in lung (Shaver's disease)
Antimony	Sb	Chemical manufacturing, solder, ammunition, cover for electrical cables, and many other uses	Antimony oxide, powdered metal, smelting fumes Stibine (SbH_3) is a very toxic gas that is formed when antimony alloys are dissolved in or washed with acid. Stibine can be released from overcharged storage batteries.	Irritates respiratory membranes, pneumoconiosis Irritates the skin (dermatitis)
Arsenic	As	Chemical manufacturing, battery plates, and cable shielding	Most forms of arsenic are toxic.	Acute symptoms: nausea, vomiting, and diarrhea, which can progress to shock and death Chronic symptoms: hair loss, discoloration or bronzing of skin, degeneration of liver and kidneys
Beryllium	Be	Metallurgy, added to copper, aluminum, magnesium, and steel	Death may result from short exposure to very low concentrations of powdered beryllium and beryllium salts.	Acute symptoms: pneumonitis, which may be fatal. Chronic symptoms: pulmonary granulomatous disease

(continued on page 148)

▶ **TABLE 7–1**
(continued)

Metal	Symbol	Uses	Toxic Forms	Toxic Exposure
Cadmium	Cd	Electroplating, metal alloys used in manufacturing, and smelting of metals	Powdered metal is toxic. Soluble compounds of metal are toxic.	Inhalation of dust or fumes causes throat dryness, cough, headache, vomiting, chest pain, pneumonitis, and possible bronchopneumonia.
			Cadmium cyanide is very poisonous (industrial use in copper bright electroplating).	Ingestion of metal of soluble compounds causes increased salivation, choking, vomiting, abdominal pain, anemia, renal dysfunction, and diarrhea.
				In Japan, people exposed to cadmium wastes in drinking water develop symptoms called itai-itai disease.
Chromium	Cr	Electroplating metal alloys used in manufacturing	Chromic acid and chromic salts are irritants to human skin and membranes.	Skin and respiratory interactions lead to ulceration.
				Oral ingestion causes severe irritation of the gastrointestinal tract, circulatory shock, and kidney damage.
Cobalt	Co	Alloys in metal manufacturing, magnets	Although cobalt is essential for life in vitamin B_{12}, only very small amounts are required. High concentrations of cobalt dust are irritants.	Inhalation of the dust may cause pulmonary disorders.
				Skin irritation is caused by cobalt dust. Ingestion of soluble cobalt solutions causes nausea and vomiting.
Copper	Cu	Chemical manufacturing alloy in metal manufacturing, wiring, and plumbing	Although copper is essential for life and the metal in trace forms is necessary, copper salts are strong irritants.	Ingesting of copper salts causes irritation of the gastrointestinal tract.

▶ **TABLE 7–1**
(continued)

Metal	Symbol	Uses	Toxic Forms	Toxic Exposure
Iron	Fe	Building materials, metallurgy, and many others	Iron oxide dust Acid dipping of iron (pickling) that contains arsenic orphosphorus can release the toxic gases arsine or phosphine.	Lung fibrosis
Lead	Pb	Storage batteries, metallurgy, paints, older plumbing connections, lead pipes, solder, as a gasoline additive, and many others	Lead dust, lead powder. Organic lead compounds (tetraethyl lead).	Toxic effects (plumbism) Acute symptoms in young children: anorexia, vomiting, and convulsions due to increased intracranial pressure. May cause permanent damage. Chronic symptoms in children: weight loss, weakness, and anemia Adults usually show chronic exposure problems after inhalation of lead dust or fumes, which causes central nervous system and gastrointestinal problems.
Manganese	Mn	Used in manufacturing steel and other metallurgy	Dust and manganese fumes Burns with intense white light when heated in air; however, it is not a fire hazard.	Inhalation of dust results in sleepiness, weakness, emotional disturbances, muscle spasticity, and paralysis.
Mercury	Hg	Also called quicksilver. Used in thermometers, mercury arc lamps, organic compounds, extraction of gold and silver from ores, and dentistry.	Elemental mercury vapors, mercury compounds, and dust with mercury	Inhaled or ingested forms are toxic. Acute symptoms: nausea, vomiting, abdominal pain, bloody diarrhea, kidney damage, and death

(continued on page 150)

▶ **TABLE 7–1**
(continued)

Metal	Symbol	Uses	Toxic Forms	Toxic Exposure
Nickel	Ni	Nickel plating, for alloys of metals, magnets, spark plugs, catalyst for hydrogenation of oils, and other organic substances	Nickel dust During high-temperature processing of nickel, nickel sulfide is reduced using coke. Nickel sulfide is a mordant in printing and blackens zinc and brass. It is toxic.	Causes dermatitis Ingestion causes nausea, vomiting, and diarrhea.
Plutonium	Pu	Radioisotope thermo-electric generator Isotope Pu-239 is used in atomic weapons. Used in nuclear power reactors	Dust is toxic and is a radiation hazard.	Concentration in bones and causes cancer
Tungsten	W	Used to increase hardness of metals. Steel manufacturing and many other uses	Dust and fumes of tungsten.	Inhalation causes respiratory problems, coughing, destruction of lung tissue, progressively leading to some reported deaths.
Uranium	U	Isotope U-234 and U-235 used in nuclear power reactors; isotope U-235 used in atomic bombs	Dust is toxic and is a radiation hazard.	Uranium salts are highly toxic. Kidney damage is observed in exposure cases. Long-term exposure to lung tissue is a cancer hazard.

Cadmium

Cadmium is an important metal with many applications in electroplating, galvanizing, and in preparing pigmentation for paints and plastics and cathode material for nickel-cadmium batteries. Cadmium is a by-product of zinc and lead mining and smelting operations, which are major sources of environmental pollution. Cadmium is readily taken up by plants, and airborne particles are small enough to be easily inhaled. Cigarette smoke is high in cadmium. Acute toxicity occurs from ingestion of contaminated foods. Cadmium is commonly used to coat metal refrigerator shelves. Using these shelves as makeshift barbecue grills can contaminate foods with cadmium. Cadmium causes pulmonary edema after inhalation. Long-term effects are obstructive pulmonary disease, emphysema, and chronic renal tubular disease, or kidney failure. It is highly carcinogenic to the prostate, lymph nodes, and lungs; it also can cause hypertension and eventual cardiovascular failure. Chelation therapy is not available for cadmium poisoning in humans.

Chromium

Sodium chromate and dichromate are used in a number of industrial processes. These are the principal substances used in producing stainless steel, chrome plating, tanning of leather, and making wood preservatives and are anticorrosive in cooking systems, boilers, and oil drilling muds. Two types of chromium plating baths are available: +3 and +6. The +6 is much more toxic than +3 and therefore should be replaced by +3 technology. Chromium is also found in ambient air from combustion of fossil fuels. While trace quantities of chromium are essential for the metabolism of carbohydrates in mammals, the major acute effect from ingested chromium is a kidney disorder, renal tubular necrosis. Cancer of the respiratory tract has been observed and is believed to be independent of dose. Skin reactions are known to occur with large exposures.

Lead

Lead is a very common toxic metal that has been detected in all phases of the environment and in all biologic systems. The metal is toxic to nearly all living things; there is no demonstrated biologic need for lead. The only major issue, then, is at what dose does lead become toxic. The principal route of exposure is ingestion with contaminated foodstuffs. Other sources include lead-based paint in old buildings, auto exhausts from leaded gasoline, industrial emissions, lead-glazed earthenware, and hand-to-mouth activity of persons living in a polluted environment.

Obviously, the factors mentioned play a large part in the disposition of lead contaminates. Target organs are the gastrointestinal, reproductive organs and systems, and, most importantly, the kidneys and the liver. Lead accumulates in the kidneys and liver causing dysfunction and eventual failure. Lead toxicity is enhanced by dietary deficiencies of calcium, iron, and, possibly, zinc. Inhalation of gasoline additives (alkyl lead compounds) is particularly hazardous because the additives are rapidly distributed to the brain, liver, and kidney by the blood. Lead poisoning can be chelated with EDTA or penicillamine. Such therapy provides moderate to fair removal of the metal from the body. However, once it is incorporated in the bone, where it has a half-life on the order of 20 years, it is difficult to ever maintain low blood levels.

Mercury

The burning of fossil fuels (coal) and the refining of petroleum products are responsible for over one-third of the mercury found in the atmosphere. Metallic mercury vaporizes, at room temperature, to form mercury vapor, which is easily inhaled and then readily diffuses across the alveolar membrane. Mercury's lipid solubility gives it an affinity for red blood cells and the central nervous system. The kidney is the primary target organ for the accumulation of mercury salts and vapor, whereas organic mercury has a greater affinity for the cortex of the brain. Mercury has the ability to cross the placenta to the fetus. Acute exposure leads to corrosive bronchitis and pneumonitis. The major effects of chronic exposure are on the central nervous system.

SOLVENTS AND VAPORS

Many tons of solvents, fluids, and dispersants are manufactured and used every year. People in commercial and residential settings are exposed to solvents. A refinery worker may be exposed to solvents on the job, then go home and paint a room, change the oil in the family car, or repair a child's toy with glue, thereby extending his exposure to solvents and their vapors. However, because different solvents have different levels of toxicity and different absorption rates, exposure cannot be easily equated with toxicity.

Properties of Solvents

Exposures to solvents are generally through the skin and by inhalation. As solvents volatilize at room temperature and evaporate and become airborne, they readily diffuse across the respiratory membrane and enter the bloodstream. Rate and depth of respiration affect the blood level concentration of solvents; distribution is controlled by heart rate. An exposure through skin contact leads to destruction of the skin tissues and, eventually, absorption into the bloodstream via penetration of the epidermis, through the hair follicles, sweat glands, and oil secretory glands. In general, solvents have the potential, at high doses, to cause cellular death. This may be caused from a single acute exposure. In the majority of cases, after removal from exposure to the solvent, recovery from the reversible effects is complete. Other general effects are manifested as changes in neurology and behavior, reaction time, psychomotor performance, and attention span. Specific toxicities are those relating to the target organs or tissues.

Aromatic Hydrocarbons

Benzene

Benzene is extensively used in the rubber industry to dissolve rubber latex. As it is a quickly evaporating solvent, it is used in printing for inks that must dry quickly; the manufacture of paints and plastics; and today as a gasoline anti-knock additive. High concentrations of benzene may be fatal because it causes depression of the central nervous system. The major toxic effect is hematopoietic and is unique to benzene in this group of aromatic hydrocarbons. Chronic exposure has been related to blood disorders such as aplastic anemia and leukemia. It is a known carcinogen.

Alkylbenzenes

Alkylbenzenes are the major feedstocks in general industry and manufacturing; tank car-size quantities are used on a daily basis. This group of compounds includes toluene (methylbenzene), ethylbenzene, cumene (isopropylbenzene), and xylene. These compounds are primarily derived from petroleum distillation and coke oven effluent. The mechanisms of action for the alkylbenzenes, with an acute exposure, resemble those of general anesthetics and can lead to CNS depression. Long-term exposure, at lower doses, is an irritant to mucous membranes. No serious long-term residual effects have been documented.

Chlorinated Aliphatic Hydrocarbons

Chloroform (CHCl$_3$)

Industrially, chloroform is a very important solvent for oils, rubber, alkaloids, and resins. The primary effects of chloroform are on the central nervous system; large doses cause liver damage, kidney damage, cardiac arrhythmias, ventricular fibrillation, and cardiac arrest.

Carbon Tetrachloride (CCl$_4$)

Carbon tetrachloride was used in fire extinguishers, for cleaning clothing, and as a solvent for oils, lacquers, and varnishes. Exposure to carbon tetrachloride in conjunction with acetaminophen or bromobenzene may result in liver necrosis.

Haloalkanes and Haloalkenes

The haloalkanes and haloalkenes are used in the plastics industry and in organic chemical synthesis. This group includes vinyl chloride, vinyl fluoride, vinyl bromide, methyl chloride-iodide and -bromide, dichloroethylene, ethyl-chloride, dichloromethane, and butane. Assessment of the risk of human exposure to these chemicals is continuing; vinyl chloride is a known carcinogen with many confirmed reports of liver cancer.

Aliphatic Alcohols

Ethyl Alcohol

Ethyl alcohol, or ethanol, has many uses in industrial settings. Humans probably have greater exposure to this poison than to any other solvent, with the exception of water. Ethanol is used daily in tank-car quantities by industry as a solvent, and it is heavily consumed every day as an intoxicating beverage by large numbers of people. The literature on the biologic medical effects of alcohol is the largest single literature in medical science.

Once ingested, ethanol is quickly absorbed from the stomach and distributed throughout the body. It rapidly reaches equilibrium with water and, therefore, every organ is affected. The major sites affected are renal, hepatic, circulatory, pulmonary, gastrointestinal, and endocrine systems. Ethanol depresses the CNS from the top down, starting with the frontal lobe and inhibitions, and working its way down to the brain stem, which controls respiratory rate and heart rate. When these latter centers are depressed, the person stops breathing and dies. In the liver, the first step in alcohol processing is to form acetaldehyde. Normally, this product is quickly degraded. However, with high ethanol consumption, there is a bottleneck in the degradatory events. The levels of acetaldehyde become toxic and cause cellular necrosis.

Methanol

Methanol, which is also known as wood alcohol, has extensive use in industry as a solvent. In humans, the fate of methanol is similar to that of ethanol. Unfortunately, the first degradation product is formaldehyde, which is a very reactive fixative or preservative. Methanol appears to have a propensity toward targeting the optic nerve, which is quickly fixed, resulting in blindness. Due to the widespread targeting ability of methanol, numerous systems are quickly adversely impacted. Ethanol competes with methanol for targets and can be used to mitigate methanol's toxic effects.

Glycols

Glycols are widely used in heat exchangers, antifreeze formulations, hydraulic fluids, and chemical intermediates. Glycols also have been used as a solvent for pharmaceuticals, food additives, cosmetics, inks, and lacquers. Taken orally, ethylene glycol (common antifreeze) is toxic to humans. Diethylene glycol is used in lacquers, makeup (cosmetics), in lubricants, as softening agents, and as a plasticizer. Inhalation of glycol vapors is a hazard, especially at increased temperatures. Propylene glycol has a low order of toxicity and is used in food products without apparent difficulty.

Glycol Ethers

Glycol ethers are used as solvents in lacquers, varnishes, resins, printing inks, textile dyes, and anti-icing additives. These solvents target the reproductive system and may cause degeneration of the testes.

Aliphatic Hydrocarbons

Aliphatic hydrocarbons are the straight-chain hydrocarbon gases present in natural gas, methane, ethane, propane, and butane. Methane and ethane are asphyxiants and do not produce systemic effects. Inhalation of the vapors of liquid aliphatic hydrocarbons results in dizziness and loss of coordination.

Hexane and Hexanone

Hexane and hexanone (methyl n-butyl ketone) are known to cause peripheral neuropathies after excessive absorption. They are used in laminating plastics and vegetable oil extraction; as solvents in glues, inks, and varnishes; and as a diluent in the production of rubber. Frequently, there is sensory loss in the hands and the feet.

Gasoline and Kerosene

Gasoline and kerosene are primarily mixtures of hydrocarbons, aliphatic hydrocarbons, and aromatic hydrocarbons. High doses may result in dizziness, coma, collapse, and death. Nonlethal doses are, generally, followed by complete recovery. (The lead in leaded gasoline is covered as the heavy metal, lead.)

Carbon Disulfide (CS_2)

Carbon disulfide is used to manufacture cellophane. This widely used solvent targets the eye and the ear and contributes to coronary heart disease.

ANIMAL TOXINS

Venomous animals are those capable of producing a poison in a secretory gland. This venom is, typically, delivered by biting or stinging. *Poisonous animals* are those whose tissues cannot be eaten due to their toxicity. Many venomous animals are poisonous but few poisonous animals are venomous.

While many poisons can be grouped by chemical or toxicological nature, this is not true of venoms. Because of the wide variations of composition and because a venom often has very selective targets, they are difficult to broadly classify. Venomous effects vary greatly depending on where it enters the body, its rates of distribution, biotransformation, and excretion.

Major families of hazardous species are:

1. *Reptiles:* Reptiles include snakes and lizards.
2. *Amphibians:* Amphibians include frogs, salamanders, and newts.
3. *Lower marine animals:* Lower marine animals include the thousands of species of protista, radiata, acoelomata, and pseudocoelomata.
4. *Mollusca-gastropods:* Gastropods are snails, slugs, cone shells, and limpets.
5. *Mollusca-bivalvia:* Bivalves are filter feeders and shellfish, like scallops, oysters, clams, and mussels.
6. *Fishes:* Fishes include stingrays, scorpion fish, weaver fish, and stone fish.
7. *Arthropoda:* As a group, arthropods inflict, by far, the most poisonings of human beings. Included in this group are spiders (jumping, black widow, violin, cobweb, and running), scorpions, centipedes and millipedes, Hymenoptera (ants, bees, wasps, and hornets), ticks and mites, and a group of others that includes mosquitoes, fleas, lice, sand- horse- & deerflies, and certain waterbugs, kissing bugs, and bedbugs.

Venomous and poisonous animals are a threat to human safety. They are a part of our environment and may be a major consideration at any work or activity site. Bites and stings from animals can have effects ranging from minute lesions of the skin to anaphylactic shock of the cardiovascular, respiratory, and nervous systems.

PLANT TOXINS

Since the earliest times, humans have been experimenting with the plant life available to them in their surroundings. They have sought to discover plants that might nourish and medicate the body. Unfortunately, many of the toxic substances in the plants caused harmful reactions. Most modern-day plant poisonings tend to occur when toddlers try to taste house plants or when preschool-age children take bites of yard plants. Adolescents may experiment with mushrooms and other plants thought to be hallucinogenic. Health food stores may carry items mislabeled or thought to be otherwise harmless but which have been treated with insecticides, weed killers, or fertilizers. Poisoning of grazing animals is also important, as it may have a significant effect on the food chain.

Plants as Toxic Agents

Most poisonous plants, regardless of their ultimate toxicological effect, will induce fluid loss through vomiting or diarrhea. They can cause reduction of lung tidal volume, which can lead to respiratory failure, even if a normal respiratory rate continues. The following sections describe some examples of plants and their hazards. Table 7–2 lists examples of some toxic plant substances.

▶ **TABLE 7–2**
Some toxic plant substances.

Common Name	Plant	Alkaloid
Poison hemlock	Coniume maculatum	Coniine
Cocaine	Erythroxylum coca	Cocaine/Ecgonine
Belladonna	Atropa belladonna	Atropine
Jimson weed	Datura stramonium	Scopolamine
Curare	Chondodendron sp.	Tubocurarine
Opium poppy	Papaver somniferum	Morphine
Ipecac	Cephaelis opecacuanha	Emetine
Ergot	Claviceps purpurea	Ergotamine
Strychnine	Strychnos nux-vomica	Strychnine
Death camas	Zigadeuus sp.	Zygacine
Peyote	Lophophora williamsii	Mescaline

Oral Cavity Irritation

The dumbcane is so named because of its immediate irritation of the tissue in the oral cavity and the oral nasal pharynx. It can cause loss of voice and poisoning of the larynx or voice box. The cultivated houseplants (Araceae) include philodendron, ceriman, malanga, and the wooded species known as skunk cabbage, green dragon, and Jack-in-the-pulpit.

Non-Diarrhetic Emesis

This group's resemblance to onions makes them ready objects for accidental poisoning and include narcissus, jonquils, daffodils, and wisterias.

Diarrhea and Emesis Response

Several plants are useful for their cathartic effects (purging), which are caused by anthraquinone. Plants containing saponins induce symptomatic responses from the use of their leaves in salads made of the pokeweed, nuts of the horse chestnut, fruit from the blue cohosh, pigeon berries, the fruit or leaf of English ivy, and the yam bean. Buttercups, baneberry, marsh marigold, pasque flower, and clematis are included.

Delayed Gastroenteritis

Colchicine-containing plants are the leaves, seeds, bulbs, or flowers of crocus or meadow saffron and tubers from the glory lily. Colchicine is a mitotic poison resulting in bone damage, hyperthermia, muscle damage, and thrombosis. Horse nettle, ground cherries, and jessamines contain solanine alkaloids that cause infectious gastroenteritis. Mistletoe contains tyramine, which causes hypertension and headache. The most toxic flowering plants are the castor bean and the rosary pea. They contain lectins, which are so deadly that a single molecule can kill a cell (necrosis).

Convulsions

Mistaken for the parsnip, the water hemlock contains convulsant toxicants. One of the most toxic plants in Great Britain, the water dropwort, has a related species in the United States.

Belladonna Alkaloid

Jimpson weed and angel trumpets contain this toxic substance. The leaves and seeds are used for their deliriant (hallucination) effects. They may also cause rapid heartbeat (tachycardia), blurred vision, and temperature increase, with the severity of response depending entirely on the individual.

Cardiovascular Disorder

The foxglove, lily-of-the-valley, oleander, and lucky nut all contain digitalis-like glycosides. The entire plants are toxic, including the smoke from their burning. Monkshood contains aconitine, which opens sodium channels in nerve axons. Green hellebores and the death chamois contain ventradine. The rhododendron, a very common household ornamental plant, contains high levels of grayanotoxins.

Muscle Tone affect

The most frequently encountered species that affect muscle tone contain nicotine or anabasine. Similar poison is produced by hemlock, laburnum, mescal or the burning bean, and cardinal flowers. Progressive stages can lead to paralysis of the respiratory tract. The Carolina jessamine and the Coyotillo, (tullidora or buckthorn) contain the toxic anthracenones.

Skin Injury

This category is extremely broad and includes plants that cause mechanical injury such as splinters, thorns, spurs, and sharp-leafed plants. Many plants cause allergic contact dermatitis, such as poison ivy and poison oak. Some plants, such as the Australia nettle, cause contact urticaria or a wheal and flare lesion of the skin. Plants from the carrot family, such as parsnips, caraway, dill, and parsley, and the citrus plants cause phytotoxicity. A number of plants cause corrosive damages similar to acid/alkali burns, such as buttercups, daphne, and wild pepper.

Mushrooms

The study of the toxicological effects of mushrooms and their treatment has led to the development of powerful evaluation tools in biochemical and neurological research. A discussion of mushrooms can be divided into their categories of response:

1. *Gut reactions:* Mushrooms whose response is gastroenteric or causes vomiting, serious abdominal pain or diarrhea.
2. *Sweat:* Mushrooms evoking sweat.
3. *Hallucinogens:* Mushrooms inducing inebriation or hallucinations without drowsiness or sleep.
4. *Delirium:* Mushrooms producing delirium associated with sleep or coma.
5. *Sensitivity:* Mushrooms eliciting a response to alcohol (sensitivity).

Mushroom intoxications may have delayed responses of six hours or more. There may be progression from those producing severe headache, to those producing severe diarrhea and emesis, to those producing excessive thirst (polydipsia) and excessive urination (polyuria) about three days after ingestion. Death has been shown to be from renal failure.

SUMMARY

Many types of plants have not been discussed, such as those producing hayfever, food allergy, or inflammation of the alveoli in the lung. There are also carcinogenic plants, teratogenic plants, and those that contain estrogen, thyroid-blocking substances, toxic amino acids, and hypoglycemic substances. In the future, continued study of plant extracts and toxins has much potential for developing new pharmacological agents. This can lead to better management, identification, and treatment of a variety of conditions.

STUDY QUESTIONS

1. How are pesticides in a unique position relative to other hazardous materials and how do most acute poisonings occur?
2. Name three classes of insecticides and for each class describe their effects and provide one or two examples.
3. What is the name of the type of pesticide that is used to kill unwanted plant life? Briefly describe the various types of such a pesticide and how they are toxic to humans.
4. Describe how fungicides are dangerous to mammals and provide several examples.
5. What is a fumigant?
6. Explain the purpose of a rodenticide and provide a list of six such chemicals.
7. Describe how metals are dangerous to mammals and provide several examples.
8. Which poison described in the text probably constitutes the largest threat of hazardous materials and poison found in every household? Write a paragraph summarizing this group.
9. Describe how animals may be toxic and provide several examples.
10. Describe how plants may be toxic and provide several examples.

8

Environmental Toxicology

COMPETENCY STATEMENTS

This chapter provides information on the basic environmental pollutants. Chemicals access the food chain through four channels: (a) directly in foods as preservatives; (b) by contaminated water, either drinking or through other plant and animal metabolism; (c) by contaminated air that we breathe; and (d) by exposure to natural and synthetic sources of radioactivity. This chapter briefly identifies examples in these categories; for the health effects of these categories, readers should refer to earlier chapters in this text, particularly Chapter 7 and also Chapter 10. After reading and studying this chapter, students should be able to meet the following objectives:

▶ Discuss in general terms the basic threat to human health from environmental pollutants.

▶ Be able to define and identify the main groups of direct food additives.

▶ Be able to define and identify the main groups of indirect food additives.

▶ List and discuss the primary reducing-type air pollutants.

▶ List and discuss the primary oxidizing-type air pollutants.

▶ Explain why water and soil are discussed together in this text and identify the principal sources of such pollutants.

▶ Identify and briefly explain the variables that influence the fate and distribution of chemicals in the environment.

▶ Develop a simple list of pesticides and give at least two examples in each category on the list.

▶ Describe the responsibilities of environmental toxicologists.

INTRODUCTION

All organisms need a niche to survive, a place where they can function and where their basic needs in life can be met, such as having shelter and finding enough of the right type of food to sustain life. Humans are no different, and since the earliest times they have been searching for food, water, shelter, and fire to keep warm and to prepare food. Humans have often been able to alter their environment to meet their needs and frequently have done so without much concern or thought about the consequences. In earlier times, even not so long ago, it was widely believed that the earth could heal itself, that there were virtually limitless natural resources, and that the earth and nature were there for human conquest.

As the human population increased, the accompanying cultivation of soils and concentration in urban areas (urbanization) resulted in the ravaging of natural resources; as a consequence, the environment was damaged. With the advent of the industrial age, the rate of pollution has increased exponentially, and humans have come to realize that environmental pollution is more than just dead fish from the dirty water or sick birds from the smoggy air: environmental pollution also adversely affects the health and safety of everyone who shares these resources. Many scientists believe that human beings have polluted their environment on a global scale and that the earth is now threatened with catastrophe.

Certainly the earth has limited resources, and humans can no longer freely exploit them without dire consequences. Humankind must clean up their mess and place their "house" in order if there are to be adequate resources for future generations who may inhabit the earth. However, the environment must be cleaned and protected from further pollution not only for the assurance of generations to come but also to improve our present-day quality of life.

In this chapter our discussion will focus on the various pollutants themselves, such as food additives and those found in the various mediums such as in the air, water, and in the soil. How different chemicals get into the food chain, or otherwise become a threat to human health and safety, may be found in environmental toxicology texts, a subject not covered here. Also included is a brief discussion of radiation and radioactive materials. All of these elements affect our health and safety and that of generations to come.

FOOD ADDITIVES AND CONTAMINANTS

Climatic conditions change with the seasons. These periodic seasonal shifts of climatic conditions have led to a pattern where there is an abundance of food during the harvest period but inadequate supplies of food during the remainder of the year. Food must, therefore, be stored to ensure that food is available year round. Methods have been developed to preserve both the harvest and game collected during peak periods.

The earliest substance added to food was sodium chloride (table salt), which is still a major preservative. The other major method was smoke, which both dehydrated and preserved the food. Spices of various types have also been used to preserve food and to disguise the flavors of decay. Today, the search for better preservatives continues. With a greater division of labor in the workforce, fewer people are needed to produce food while more and more consume food. In part, this is possible because technology allows fewer to grow and produce more food. Efforts to increase food production, storage, and use have created a corresponding increase in environmental pollution.

Processed foods now represent over 50% of the American diet. Trends, such as the demand for "ready-to-eat" foods, an interest in ethnic cuisine, and snacks have led to a tremendous market (demand) for low-cost, instant, year-round supplies of seasonal food. Technology must now consider how to meet these demands in a manner that is friendly both to humans and to the environment.

Definition

It is our government's duty to protect public health and safety. To this end, the government has enacted a number of laws and regulations. To ensure that everyone knows what is meant by the regulations in the Food, Drug and Cosmetic Acts and the Meat Inspection Act, a number of terms have been explicitly defined. Some definitions are associated with technical use, such as for substances added during the production, processing, and storage of food. Other definitions help consumers evaluate the benefits and risks associated with food additives. Finally, some definitions standardize labels to allow the public to analyze container contents. The following are definitions of some of the key food additives.

Direct Food Additives

The FDA has estimated that approximately 95% (by weight) of the total food additives are those substances intentionally added to food: sucrose, corn syrup, dextrose, salt, black pepper, caramel, carbon dioxide, citric acid, starch, sodium bicarbonate, yeasts, and yellow mustard. It is interesting that 97% of this total is represented by sugars (sucrose, corn syrup, dextrose) and salt.

Main Groups

The main groups and the various categories of direct food additives are as follows:

1. *Processing aids:* Processing aids are emulsifiers, pH control agents, enzymes, thickeners, lubricants, and leavening agents.
2. *Texturing agents:* Texturing agents give food a desirable consistency and texture.
3. *Preservatives:* Preservatives are used to decrease the rate of degradation during processing and storage.
4. *Flavoring and appearance agents:* Flavoring and appearance agents improve the palatability of food and are sweeteners, waxes, and surfacing agents.
5. *Nutritional supplements:* Nutritional supplements are added to replace substances lost during processing or to supplement existing levels of nutrients. These are typically vitamins and trace amounts of minerals.

6. *Color:* Color is used to enhance, alter, and produce more appealing foods. Agents include both natural and synthetic colors. This practice dates back to before the 4th century B.C. Color additives receive the majority of public, scientific, and regulatory attention.

Indirect Food Additives

This group consists of the class of substances not naturally a constituent of food. They have not been added to food for any technical purpose. They come from a variety of sources, such as the environment in which the food is produced, processed, or stored and from subsequent packaging and storage for the market. Potential indirect sources are:

1. *Drugs:* Drugs are used in treating or dosing of food-producing animals. These bioaccumulated and are absorbed by the consumer. They may produce chronic health effects or may be carcinogenic.
2. *Pesticides:* Pesticides include all classes of insecticides, fungicides, and herbicides that are used or found in the environment.
3. *Hazardous compounds:* Toxic and radioactive compounds, both natural and synthetic, are loose in the environment. These settle on the ground or enter the water supply. Bioaccumulation occurs and the toxins concentrate up in the food chain.
4. *Microorganisms:* Microorganisms are everywhere and grow whenever given a chance. These can enter the food chain and make people sick. Some bacteria, like Botulum, produce toxins that remain toxic even after the organism is dead.
5. *Residuals:* Residuals include the leftovers from normal processing and foreign objects accidentally added during processing.
6. *Containers:* Containers are manufactured and often have substances on the surfaces or within the material that can leach into the container contents. Other packaging and storage sources are labeling and stamping materials, migrants of microorganisms from packaging materials, and toxic chemicals from external sources.

Food Contaminants

Food substances are comprised of a complex mixture of a vast diversity of biological substances and components of organisms. Foods may be contaminated with molds, bacteria, parasites, insect parts, and the feces and hair of rodents and other animals. Contaminants may find their way into the food chain through irrigation water, general environmental pollution, or chemical spills.

AIR POLLUTANTS

With increasing urbanization, the problems associated with humans' need to prepare food and keep warm have been compounded by increases in population and by industrialization. The factories built to produce the raw materials to manufacture the necessities of our everyday lives produce large amounts of waste. Historically, one of the trade-offs for this economic development has been smokestacks belching pollution far up into the air. The entire food chain is affected as winds blow this pollution around the world and rain leaches contaminants from the air to the ground and waterways.

As people moved into the suburbs to get away from the dirty city air, pollution was compounded by the proliferation of the automobile. The internal combustion engine is responsible for creating "smog" from exhaust gases and spewing the lead from gasoline additives. Large amounts of chlorofluorocarbons are used by our society. Their accidental and intentional release into the atmosphere is endangering our environment by contributing to smog formation and, it is hypothesized, destroying the ozone layer. The magnitude of the damage is being closely monitored. It is believed that the opening hole in the ozone layer in the Antarctic region is due to the chlorofluorocarbons that humans have created and released into the atmosphere. Our self interest, to keep ourselves warm, power our labor-saving devices and lights, produce the gadgets we need to make life more "enjoyable," and to take ourselves to and from work, has damaged the atmosphere and therefore jeopardized the ecology of the global environment.

Reducing-Type Pollution

Smoke resulting from the incomplete combustion of coal, along with fog and cool temperatures, is termed *reducing-type air pollution*. The burning of fossil fuels and the smelting of metals emit a variety of particles and gases, such as fly ash containing heavy metals; carbon, sulfur, and nitrogen oxides; and radioactive materials, into the atmosphere. In the atmosphere, sulfur dioxide is mixed with water and other reactants to produce sulfuric acid (H_2SO_4). Nitrogen oxides likewise are converted into nitrogen acids. These acids are absorbed on the surface of minute particles of airborne metals and are transported great distances by the winds. They pose a direct health hazard and contribute to acid rain. Acid rain is considered by many to be the major ecological problem in our northern hemisphere. Smog and acid rain have been directly linked to the killing of lakes and the destruction of forests.

Sulfur Dioxide (SO_2)

The EPA has acknowledged it is of little significance to analyze SO_2 without also considering the coexisting particulate matter. However, studies have shown that SO_2 produces damage in the tracheal ciliated epithelial cells and near destruction of the associated goblet cells. This damage results in a pathology similar to acute bronchitis, with accompanying difficulty in breathing. Asthmatics are particularly sensitive to even low levels of sulfur dioxide. In the presence of a nitrogen source, zinc and vanadium convert SO_2 to sulfuric acid, ammonium, sulfate, and ammonium bisulfate.

Sulfuric Acid (H_2SO_4)

This strong acid promotes laryngeal spasm and bronchospasm. Depending on concentration and individual resistance, it can be lethal. Sulfuric acid also produces parenchymal lung damage.

Sulfate Salts

Sulfate salts and other aerosol particulates interact with SO_2 and H_2SO_4 and increase the health risks associated with each. Of most concern are the metals manganese, iron, and vanadium.

Carbon Particulates

Incomplete combustion of fossil fuels produces soot (carbon particulates), which drift far and wide in the atmosphere. Inhalation of these particulates leads to their accumulation in the lung and associated respiration problems. Frequently, toxic metals and carcinogenic organics adsorb onto soot and are transported deep into the lung.

Oxidizing-Type Pollution

Oxidizing-type pollution is the second type of air pollution and is the result of the atmospheric reaction to the products of internal combustion engine exhausts. The presence of incompletely combusted hydrocarbons is of concern because they enter into chemical reactions that lead to the formation of photochemical smog.

Ozone (O_3)

Ozone is the oxidant found in the largest amounts in polluted atmospheres. Ozone (O_3) is a deep lung irritant capable of causing asphyxiation from edema. Ozone is created from oxygen (O_2) miles above the earth's surface when ultraviolet light (UV) is absorbed by oxygen. If no additional mechanism were involved, most of this ozone would be broken down by nitrate (NO), but the hydrocarbons present (especially olefins) are attacked by oxygen. These reactions produce oxidized compounds, which then react with NO to produce NO_2, the most efficient absorber of the wavelengths of UV light that reach the earth's surface.

Normally, these reactions form a fully cyclic reaction, which is not a problem. However, we have upset the balance so that more and more NO_2 is being produced, which absorbs more of the UV radiation so that more ozone (O_3) is created from the oxygen, which depletes the NO levels. The result is a decrease in beneficial radiation reaching the earth's surface and an increase in the production of formaldehyde, acrolein, and peroxyacetyl nitrate (PAN), a potent eye irritant.

Exposure to ozone produces alterations in respiration and an increase in air flow resistance. It also increases the lungs' sensitivity to bronchioconstrictive agents such as acetylcholine, histamine, and allergens. This means a greater susceptibility to bacteria and other microorganisms.

Nitrogen Dioxide (NO_2)

Nitrogen dioxide, like ozone, is a deep lung irritant capable of producing pulmonary edema. NO_2 causes the lungs to become emphysemic due to extreme vascular dilatation, collapse of the alveoli, and lymphocyte infiltration.

Aldehydes

Aldehydes, especially formaldehyde and acrolein, are formed as reaction products in the photooxidation of hydrocarbons. Aldehydes are responsible for the odor and eye irritation of smog. They are primary irritants that are soluble in water and therefore can invade the mucous membranes of the eye, nose, and upper respiratory tract. An irritant aerosol is formed by the submicron particles of a sodium chloride aerosol attaching the aldehydes. This potentiates the effects of these pollutants on humans. Acrolein is much more potent than formaldehyde; concentrations of just one part per million (ppm) are sufficient to irritate the respiratory tract.

Carbon Monoxide

Carbon monoxide is classed as a chemical asphyxiant because it denies attachment of oxygen to hemoglobin molecules in the blood. Combined with smog, carbon monoxide has produced carcinomas in mice from short-term, high-level exposure. In chronic cases, it injures the airways, stimulates development of lesions in the bronchial tree, and damages the alveolar epithelial cells.

Conclusions

The Clean Air Act established guidelines and legal steps for the attainment of air quality standards. Cleaning up the air we breathe and establishing tighter limitations on acceptable levels of emissions are critical to maintaining our health and safety.

WATER AND SOIL POLLUTANTS

Although they are actually two separate systems, water and soil are so intimately linked that it is difficult to speak of one and not of the other. Understand that soil particles are suspended in water and that subterranean water systems, the aquifers, are suspended in the soil. Because of this close interface between the two systems, contamination of one affects the other.

Three quarters of the earth's surface is covered with water. Ground, soil, and rock, unless it has been paved over, covers the rest of the earth. Bodies of water are recipients of run-off from rainwater, agricultural and urban areas, eroded rock and soils, and the products of domestic and industrial sewage systems. Water can, at an expense, be purified. The greater the purity, the more expensive the treatment. Soil, too, can be cleaned, but the concept is different.

Water (H_2O) is a specific chemical with unique properties. Soil is composed of a vast mix of organic and inorganic constituents. Soil may take the form of silt, sand, and clay. Typically, these are coated with many organic constituents both living and dead. The behavior of soil largely depends on the size of its particles. The size, too, varies considerably, from clay (at less than 0.002 mm) to small pebbles (coarse sand about 2.0 mm diameter). The sediment, which is the end product of soil erosion, is by far the single greatest pollutant of surface water. However, both agricultural and industrial effluents have vast local effects and greatly aggravate the problem. Sediment is the principal carrier of pollutants found in water.

Sources of Chemicals in the Environment

Industrial Processes

Industrial processes produce a vast array of by-products and waste products. In the past, disposal has been to the environment with a subsequent pollution and contamination of soil and water. While appropriate use of technology can make a difference, it does not prevent or diminish improper handling and disposal of hazardous materials. The results of accidents, ignorance, and criminal behavior have a detrimental effect on our environment.

Agricultural

Agriculture is one of the largest users, and abusers, of chemicals. Overfertilization leads to water run-off pollution, irrigation leads to salt pollution of the soil, pesticides are applied directly to the soil where they can persist for many years, accumulate in the food chain, and eventually, leach or run off into nature's water systems.

Domestic and Urban Use

Although the amounts of water and chemicals we use may seem small when compared to that used by industry, the numbers become large when our individual use is multiplied by hundreds of millions. Household disposal, though small as individual quantities, collectively accounts for a tremendous volume as home pesticides, fertilizers, detergents, paints, solvents, and lawn care products become concentrated in sewage systems, storm water systems, and landfill operations.

Metals, Minerals, and Plant or Animal Toxins

Metals, minerals, and a variety of plant and animal toxins are found naturally in the environment. These appear as components of water and soil systems. Human activities frequently disturb such materials at their sources, increasing their assessibility to the environment and allowing concentrations detrimental to the support of life. Furthermore, the interaction of synthetic materials, chemicals, and wastes with naturally occurring agents and organisms creates a variety of environmental pollution such as mine leachates.

Transport, Mobility, and Disposition of Pollutants

The fate and distribution of chemicals in the environment depend on a number of variables with multiple interactions. Some pertinent factors include the following:

1. *Water solubility:* Water solubility is an important consideration when trying to predict the behavior of water-based systems.
2. *Latent heat:* Latent heat becomes very important in, for example, purification processes. The evaporation of water and its recondensation transport vast amounts of heat around the world.
3. *Adsorption:* Adsorption of chemicals on particulate matter is an important mechanism by which chemicals are removed from solution and transported through the environment.
4. *Vaporization:* Water vaporization from bodies of water, soil, and plant surfaces is a major transport process; a chemical with both high solubility and high vapor pressure would be very volatile from aqueous solution.
5. *Partitioning:* Partitioning is the phenomenon of dispersion throughout the various segments of the environment. It can be represented as the ratios that an agent, initially released from a point source, can eventually be divided into such mediums as the air, the soil, water or into plants and animal tissues.
6. *Bioaccumulation:* Bioaccumulation is an active process, that is, it requires the expenditure of energy. At each step in the food chain, the concentration of, for example, a toxin, may increase ten to one hundred times. At some point the accumulated concentration is great enough to cause problems. Examples are the

effects of the accumulation of DDT on bird eggs and the problem with ciguatera poisoning from eating certain fish. Bioaccumulation is also different from the other passive processes in that toxins accumulate in organisms rather than diffuse into the environment.

7. *Degradation:* Degradation causes a true "disappearance" of a substance by chemical, biochemical, and photochemical reactions.
8. *Chemodynamics:* Chemodynamics is the behavior of chemicals in the environment.

Pesticides

Pesticides, which persist in the environment past the time needed to perform their designed function, have become a very important ecological and environmental problem. *Persistence times* reflect the time required for 75% to 100% of a chemical's disappearance from the site of application. Nonpersistent pesticides have values of 1 to 12 weeks. Moderately persistent have values of 1 to 18 months, and persistent, from 2 to 5 years.

Persistent Pesticides

The most persistent pesticides are the chlorinated hydrocarbon insecticides such as DDT, TDE, methoxychlor, aldrin, dieldrin, heptachlor, chlordane, mirex, kepone, BHC, lindane, and toxaphene. Additionally, this group includes the cationic herbicides (paraquat and diquat), which are used in conservation tilling.

Moderately Persistent Pesticides

The moderately persistent pesticides include the triazine herbicides, and altrazine is the compound of greatest concern. The phenylurea herbicides are also in this group. Listed in descending order of persistence are fenuron, monuron, diuron, linuron, monolinuron, fluometuron, metobromuron, norea, siduron, neubron, and chloroxuron. This group also includes the dinitroanilines such as trifluralin, oryzalin, pendimethalin, and related materials.

Nonpersistent Pesticides

Some pesticides that do not persist include the following:

1. The phenoxyl and related acidic herbicides, which possess carboxyl groups, and ionize in aqueous solution. Examples are 2,4–D, dalapon, trichloracetic acid, chloramben, dicamba, diclobenil, and dinitro-o-cresol (DNOC).
2. Phenylcarbamate and carbamilate herbicides include propham, chlorpropham, barban, terbutol, and dichlormate.
3. Ethylenebisdithiocarbamate fungicides and their metal derivatives are represented by ferbam and ziram.
4. Synthetic pyrethroids are a good trade-off between persistence versus residual activity, though they are toxic to fish.
5. Organophosphorus and carbamate insecticides have generally replaced the hydrocarbon variety and include aldicarb, carbaryl, diazinon, parathion, zinophos, malathion, supracide, dimethoate, and chlorpyrifos.

Nonpesticidal Organic Chemicals

Low Molecular-Weight Hydrocarbons

Low molecular-weight, halogenated hydrocarbons are produced because of the availability of halogens due to their use in water purification as a chlorinator. Concerns about environmental pollution focus on chloroform, bromo- and dibromochloromethane, bromoform, carbon tetrachloride, and dichloroethane.

Aromatic Halogenated Hydrocarbons

The aromatic halogenated hydrocarbons include polychlorinated biphenyls (PCBs), chlorophenols of pentachlorophenol and hexachlorophene, and 2,3,7,8 tetrachlorodibenzo-p-dioxin (TCDD), a contaminant in an herbicide known as Agent Orange, which has been found to be extremely toxic to animals.

Phthalate Ester Plasticizers

Plasticizers are used in virtually every major product category, including construction, automotive, household products, apparel, toys, and packaging. The two most abundant are hexylphthalate (DEHP) and di-n-butylphthalate (DBP). These chemicals occur widely in the environment but they do have a low acute toxicity. Yet, their tendency to accumulate over extended periods is definitely a cause for concern.

Metals

The most environmentally important metals are the following:

1. *Mercury:* Mercury is a concern mostly in its organic, methyl compounds.
2. *Cadmium:* Deposits of cadmium are found naturally as sulfides with zinc, copper, and lead as by-products of their smelting process.
3. *Lead:* Lead is now ubiquitous in our environment.
4. *Arsenic:* Arsenic is widely spread in the environment as a result of smelting, coal burning, and use of pesticides.
5. *Selenium:* Selenium, now thought by some to have a beneficial anticarcinogenic effect is distributed by smelting, burning of fuels, glass and ceramic pigmentation, and uranium mining.

Inorganic Ions

Nitrates

Nitrates are widely used in agriculture and technology and have upset the nitrogen cycle. Nitrate fertilizers have high water solubility and great mobility in both soil and water. They are of concern because of their pollution of fresh water and their easy conversion to nitrites, which creates oxygen transport difficulties in humans.

Phosphates

Phosphates are essential to life processes, although they are not directly related to human health problems. Phosphates are a major cause of the eutrophication process in lakes and ponds. The sources are mainly detergents, insecticides, and fertilizers. Phosphates and nitrates stimulate algae growth which, in turn, depletes the oxygen content of the deep waters.

Fluorides

Fluorides are popular in the treatment of drinking water to prevent tooth decay. Acute effects are not seen from normal levels, but chronic high levels can result in crippling skeletal fluorosis.

Asbestos

Asbestos are fibrous minerals that present occupational inhalation hazards and, technically, become air pollutants. While occupational exposure to inhaled asbestos fiber has been well quantified, the effects in soil and water are still under examination.

Asbestos fibers have superior thermal and electric insulating capabilities and can be easily woven into fabrics and cloths. The minerals that comprise asbestos are serpentine, chrysotile, actinolite, amosite, anthophyllite, crocidolite, and tremolite. Chrysotile accounts for about 95% of the asbestos used in the production of products. Asbestos is a very heterogeneous material and its detection is difficult. Asbestos is found ubiquitously in the environment. Contamination of drinking water by asbestos has been identified.

Chemical Waste Disposal

Sources of hazardous wastes include homes and industry, government, agriculture, laboratories, universities, hospitals, and power plants. Wastes take the form of solids, sludges, liquids, and gases and are classified as ignitable, corrosive, toxic, explosive, biologic, and radioactive. These materials are directly hazardous to health and safety, and they contaminate the soils and water systems. A concerted, well-coordinated, cooperative attack to prevent and cure these serious environmental problems must be conducted worldwide.

Conclusion

The traditional view of the environment as a self-purifying infinite resource as embodied in the phrase "balance of nature" is outmoded and oversimplified. The environment is composed of many interacting systems and, in this diversity, humans must understand and manage the effects of these interactions. The toxins we generate negatively affect ourselves and our environment.

SUMMARY

▶ Toxins in the environment may be in the form of food additives, criteria air pollutants, industrial and agricultural chemicals and pesticides, and radiological hazards. These are found in the air, water, and soils.

▶ Modern science views the existence of the living world as depending on the flow of energy and on the continuous cycle of materials from medium to medium in the biogeochemical cycle. Materials dumped into the environment in one location diffuse and permeate to the whole world.

▶ Environmental toxicologists study how pollutants enter and are cycled by the earth. They also determine what chemicals, materials, and substances exist in the environment, identify which are a result of disaster or human intrusion, and identify their impact. They are concerned with the effects these chemicals may have on ecological health and safety.

▶ Humans have interfered with and manipulated the environment for thousands of years. Today, few people have a personal environmental ethic. The aesthetics of a lovely view or a nice day are symptoms of a pristine environment and healthy ecology that are seen too rarely today. Modern civilization has exploited these resources with the view that modern technology can rescue us from our environmental problems. There are no easy solutions. Each of us makes an impact and, therefore, must work to harmonize with nature. Historically, institutions have been slow to respond to problems. Even Scripture (Genesis 1:26), which gave humans dominance over nature and permitted great exploitation, needs to further evolve into a concept of humans *and* nature rather than humans *in* nature.

STUDY QUESTIONS

1. What is the basic threat to human health from environmental pollutants?
2. What are food additives and what are the main groups of direct food additives?
3. What are the main groups of indirect food additives? Give examples in your answer.
4. What are primary reducing-type air pollutants? Give examples in your answer.
5. What are primary oxidizing-type air pollutants? Give examples in your answer.
6. Why are water and soil discussed together in this text? Identify the principal sources of such pollutants.
7. What are the different variables that influence the fate and distribution of chemicals in the environment?
8. Develop a simple list of pesticides and give at least two examples in each category on the list.
9. What is the basic role and responsibility of the environmental toxicologist?

9

Radiation Health Effects

COMPETENCY STATEMENTS

This chapter discusses the scope, use, and hazards of radioactivity used in industry, medicine, and the nuclear industry. After reading and studying this chapter, students should be able to meet the following objectives:

▶ Describe the units of the roentgen, curie, becquerel, rad, gray, rem, and sievert.

▶ Describe the acute and chronic effects of ionizing radiation.

▶ Describe the concept of natural background radiation.

▶ Discuss the radiation protection standards and define the NCRP and ICRP and their functions.

▶ Give an overview of regulatory agencies that deal with radioactive materials.

▶ List the sources of ionizing radiation that are found in the environment and describe the various biological responses and different factors that influence those responses in humans.

INTRODUCTION

Radioactive materials and sources are commonly used in a large variety of medical, commercial, government, laboratory, and industrial applications. The hazards to workers from exposure to these sources must be recognized and appreciated. Some applications of radioisotopes include industrial radiography, borehole logging, radiation gauging, smoke detectors, and luminous materials. Because most of these radioisotopes are packaged as enclosed, sealed, or encapsuled, they are therefore not in a form free to migrate. However, workers may receive exposure during shipments, maintenance, or disposal of these sealed sources. In the medical field, radioisotopes are used in diagnosis of diseases, therapy for diseases (cancer treatments), and for heart pacemaker batteries. The medical use of radioisotopes includes both sealed and unsealed sources. Another radiation source is the X-ray tube, used for X-ray fluoroscopy, and image production. Health care professionals and patients must be protected from these occupational exposures. Aprons with lead-shielding reduce the exposure of sensitive body parts such as the gonads to ionizing radiation. Radioactive tracers used in nuclear medicine expose the patient, hospital personnel, transport personnel, and radioactive waste worker. It is important to minimize the doses. There is an inherent risk in the use of radiation. However, it is very helpful in diagnosing and treating diseases.

Occupational exposures are known to occur with some frequency in the nuclear power industry. The process of mining the uranium ore, processing it for the reactor, and then removing the spent nuclear fuel for disposal is called the *nuclear fuel cycle.* Workers are exposed to dust with radon and radon daughter products in the process of mining and milling of uranium ores. Nuclear reactor operators are routinely exposed, although, on average, at doses much smaller than those due to medical or industrial exposures and occasionally larger exposures in case of an accident or system failure. In the Chernobyl accident, firefighters received very high lethal doses of ionizing radiation. In the transport of nuclear fuels and high-level radioactive waste, occupational exposure does occur. The radioactive waste disposal process exposes workers to radiation in the burying of the waste in trenches or underground storage vaults (Table 9–1).

RADIATION AND RADIOACTIVE MATERIALS

Our increasing use of radiation by people in the modern world requires that we devote a separate section to information about the effects of radiation. Radiation has both natural and synthetic sources in the environment.

Basic Physical Concepts

The nuclei of certain unstable elements spontaneously decay to form different nuclei. This process, called *radioactivity,* causes a release of energy and often involves the formation of one or more small particles. Some atoms are naturally unstable. Others can be made unstable by bombarding them with subatomic particles. Still others may be stable, but their isotopes may not be stable. When the small particles (emissions) are released, they may pass through other matter (or tissues), depositing energy and leaving a trail of ions and molecular debris.

Radioactive decay is an atomic transmutational phenomenon. This means that a mass of radionuclide will eventually become drained of available atoms. The half-life is the time required for the number of atoms present in a particular sample to decrease by half of their original number. The rate of disintegration is an intrinsic property of

▶ **TABLE 9–1**
Examples of radionuclides workers may encounter in industry, medicine, and the nuclear fuel cycle.

Industry Application	Radioisotope		
Industrial radiography	Cs-137	Co-60	
Borehole logging	Cs-137	Co-60	
	Pu-Be	Am-Be	
Radiation gauges	Ce-144	Cs-137	Co-60
Smoke detectors	Am-241		

Medical Application	Radioisotope
Bone scan	Tc-99m
Thyroid uptake	I-131
Blood volume	I-125
Red blood cell survival	Cr-Sl
Bone cancer metastases (spreading)	P-32
Vitamin B_{12} absorption studies	Co-60

Nuclear Fuel Cycle	Radiation Hazards
Mining and milling	Radon
	Dust with uranium, thorium, and radium
Enrichment of fuel	UF_6 gas
Reactor operations	Gas-I, Kr, Xe
	Liquid-activated corrosion products
	Solid waste—spent fuel-filter cartridges

the radionuclide. The number of decays for any particular mass of radioactive material is directly proportional to the number of atoms initially present, and thus, when half of the atoms have decayed, the disintegration per second (the activity) will have decreased by 50%, then by 75% over the next half-life, then by 87.5% over the next, then by 93.75%, and so on. If the half-life of the radioisotope is known, the amount remaining after a given period of time can be calculated. These calculations are exactly the same as when we used the half-life concept to discuss the dose of a toxin in our bodies.

Measurement of Radiation

The *roentgen* (R) is a unit of radiation exposure and is related to the ability of X rays to ionize air. The process (ionization) results in the formation of plus- and minus-charged particles (ions) that are a result of the transfer of X-ray energy. This very small ionization (.000258 units plus or minus) charge per kilogram of air is used as the basis for standardizing equipment that detects radiation. Since the exposure, dose rate, is usually less than one roentgen per hour (1 R/hr), subdivisions are used, with a subdivision of one thousandth of an R/hr being called a milli R/hr and a subdivision of one millionth of an R being called a micro R/hr.

The activity of a radioactive sample is often expressed in curies, where 1 curie is 37,000,000,000 disintegrations per second. Since a curie is a large amount of activity, subdivisions of one thousandth of a curie (a millicurie) and a subdivision of one millionth of a curie (a microcurie) are used. The curie was originally defined as the number of disintegrations per second occurring in 1 gram of radium 226. The curie is being replaced by the international unit of the becquerel (Bq) as the unit of activity, where 1 Bq is equal to 1 disintegration per second.

For workers who deal with radioactive materials or radiation-producing machines (X rays, X-ray tubes, synchrotons, and particle accelerators), the absorbed dose is an important concept. You may be exposed to radiation, but not all of the radiation hits your body and not all radiation is absorbed by the cells of your body. The rad is the unit of absorbed dose. In the case of X rays and gamma rays, an exposure of 1 R (roentgen) results in an absorbed dose in soft body tissues equal to about 1 rad. Officially, the rad has been replaced as an international unit by the gray (Gy). The internationally used and recognized unit of the gray corresponds to an energy absorption of 1 joule/kg, with 1 gray equalling 100 rads.

All types of radiation can produce measurable effects on living cells. Some radiation does greater damage to cells and is more effective in harming the cells. Thus, to determine the relative biological effectiveness (RBE) for radiation protection purposes, doctors and scientists consider the following factors: dose and type of radiation, type of cells involved, and damage the radiation can cause internally or externally (called the quality factor). From these factors radiation protection officers calculate a dose equivalent for various organs or whole body in units called a sievert (Sv) or units of rem (roentgen equivalent man). One sievert corresponds to an energy absorption in body tissues of 1 joule/kg and 1 sievert is equal to 100 rem (Table 9–2).

Alpha Particles

An *alpha particle* is a helium atom stripped of its two electrons, leaving it with just two protons and two neutrons. Without the electrons, the particle has a positive charge of +2, shown as He+2. Whether it is photographic film, lead shielding, or body tissue, the alpha particle interacts with all matter. In its path it deposits energy and can create hundreds of thousands of ions along its way before taking on two more

▶ TABLE 9–2
Units that describe
radioactivity.

Quantity	Units
Activity—disintegration of radioactivity per second	curie 3.7×10^{10} dis/sec becquerel 1 dis/sec
Exposure—charge liberated by ionizing radiation per unit of air	roentgen (R)
Absorbed Dose—energy absorbed from ionizing radiation	rad (older term) gray (Gy) 1 Gy= 100 rad
Dose Equivalent—type of radiation energy absorbed from ionization radiation per tissue mass per unit time	rem (older term-radiation equivalent to man) sievert (Sv) 1 Sv = 100 rem

electrons to become a neutral helium atom again. Because of these energy deposition properties, alpha particle beams have become important medical treatment modalities for tumors and cancer.

For naturally occurring radioactive materials, the penetration of alpha particles is only about 4 centimeters (1.6 inches) in air or 1/1000th of a centimeter (0.0004 inch) in human skin. These alpha particles are, therefore, unable to penetrate human skin, but if taken internally, can do massive damage. Alpha emitters are never used in nuclear medicine. A sheet of paper is adequate protection from He+2.

Beta Particles

Beta particles, like alpha particles, interact with any matter in their path. However, the beta particle is much smaller and many times faster when ejected from a decaying nucleus. They also produce less ionization than alpha particles do. The beta particle has about 100 times the penetrating power of an alpha particle. Only 0.5 centimeter (0.2 inch) of body tissue, 1 centimeter (0.4 inch) of aluminum, and 25 centimeters (9.8 inches) of wood are required as an adequate stopping barrier. The lower ionization level with higher penetration capability makes the beta more suitable for radiation therapy. Natural emitters such as calcium–46, iron–59, cobalt–60, and iodine–131 are used extensively.

Gamma Rays

Gamma rays are usually emitted whenever either alpha or beta particles are created. Though they carry no charge, gamma rays travel many times the distance of beta particles and still produce ionization as they pass through matter. Gamma rays are easily detectable. All beta emitters also emit gamma rays. These include chromium–51, arsenic–74, technetium–99, and gold–198, which are used in diagnostic nuclear medicine. Gamma rays can penetrate human tissue 50 centimeters (19.7 inches), and air 400 meters (437 yards). Lead plates over 3 centimeters (1.2 inches) in thickness are generally adequate protection from naturally occurring gamma ray sources.

X Rays

X rays are a form of ionizing electromagnetic radiation similar to gamma rays. They are produced by a high voltage X-ray tube. A source of electrons shoots the electrons at a collecting terminal, or target, within a vacuum tube. When the electrons strike a suitable target, such as tungsten, they "knock off" X rays, which are then emitted from the X-ray tube.

Biological Effects of Ionizing Radiation

If a person receives a single, very large, whole-body dose of radiation, a number of vital cells and organs are damaged quickly. Scientific knowledge from the atomic bombs of Hiroshima and Nagasaki and other radiation events allows us to develop a pattern of biological effects known as acute radiation syndrome.

The condition of acute radiation syndrome has four stages. In the first, or prodromal, period the person will feel tired, nauseous, lose their appetite, and have some sweating. The second, or latent, stage lasts from two days to two or three weeks and is characterized by a feeling that all is well. The third, or manifest illness, stage appears as damage to blood-producing cells in bone marrow, and the patient suffers bleeding and infection. At high radiation doses intestinal problems will also be present. In addition, the person will have fever, experience hair loss, and have neurological problems, such as with perception. If the person survives, a fourth, or recovery, stage takes place and lasts several weeks.

Biological Response to Exposure

Many factors, both extrinsic and intrinsic, modify the response of living organisms to a given dose of radiation. Extrinsic factors include the dose rate, the quality of radiation, and the portion of the body exposed. The biologic factors include age, gender, oxygen tension, and metabolic status of the organism.

External Exposure

The early effects of ionizing radiation are death of cells. Those cells undergoing mitosis are the most sensitive. These include hematopoietic, intestinal, spermatogonia, ovarian follicles, epidermal, and lymphocytes. A second, and slightly less sensitive, group includes bone marrow and secondary differentiated oocytes (ova) and spermatocytes. The third class is intermediate in sensitivity and includes endothelial cells, fibroblasts (clotting), and mesenchymal (digestive) cells. Virtually all cells are sensitive, but those that are the most resistant are the long-lived, usually nondividing cells, such as cells of the liver; those that do not divide (amitotic), neurons, and WBCs; some muscle cells; and erythrocytes (red blood corpuscles).

Acute whole-body radiation produces anorexia, nausea, vomiting, diarrhea, apathy, tachycardia, fever, and headaches. For the LD50 and 30-day doses, the reader is referred to appropriate value charts. Acute partial-body exposures usually demonstrate effects to the skin and, especially, the eye, salivary glands, and the chemical, mechanical, and photoreceptors of the senses. Cumulative effects of high or low levels of ionizing radiation are the subject of much study. The hazards of radiation are offset by the benefits of therapy.

Internal Emitters

Internal emitters are radioactive materials that gain access to the body. The activity depends on the half-lives of the material, the quantity, the tissue weight, and the type of energy or particles released. The primary routes of entry for internal emitters include (a) ingestion and (b) inhalation. This information identifies the principal levels of protection, which would include respiratory protection and hygiene and safety procedures. These procedures include no smoking, drinking, eating, or chewing gum to avoid any hand-to-mouth contact in areas of potential exposure.

Chronic or Long-Term Effects of Radiation

Scientific knowledge of long-term radiation effects has come from studies of uranium miners, radium watch dial painters, the use of X rays by physicians and dentists early in this century, and populations exposed to nuclear weapons and fallout from weapons tests. In the early part of this century, doctors noticed an increase in lung cancer and lung fibrosis in uranium miners. The workers' long-term (chronic)

▶ TABLE 9–3
Acute radiation syndrome for
gamma radiation.

Dose (rad)	Symptoms
0–30	None and no detectable immediate effect
30–100	Prodromal symptoms, mild nausea, bone marrow damaged, decrease in red blood cells, decrease in white blood cells and platelets
100–300	Mild to severe nausea, loss of appetite, infections, severe bone marrow damage, recovery possible but not assured.
300–600	Severe effects, bleeding, infection, diarrhea, loss of hair, temporary sterility, fatalities in over half of cases
over 600	Death

exposure to radiation was from breathing radon gas and radon decay products in the mine air.

Another example of long-term industrial exposure to radiation came from the need to make watch dials visible in the dark. Workers painted the numbers on the dials with radium paint that would glow in the dark. As these workers tipped their paint brushes with their tongues, they swallowed radium. This practice was stopped in the 1920s. The isotopes of radium became deposited in the bones of these workers, and elevated levels of bone cancers were observed in the workers in later years. The information gained from studying a registry of these radium dial painters allowed for a baseline of radiation protection standards to be established.

Early in this century physicians and dentists received large doses of X rays, a side effect from exposing patients for treatment. Follow-up studies of thousands of doctors have shown that chronic low-dose exposure to X rays increases the development of leukemia. In Japan, survivors of exposure to atomic bomb radiation showed a marked incidence of leukemia.

Many of the long-term effects of radiation develop after years; even though damage had occurred, the workers had no manifest symptoms at the time of the radiation exposure. Bone, bone marrow (leukemia), lung, thyroid gland, and skin cancers have been observed and recorded. In many cases, ten years or more passed before the cancer was detected. Life-span shortening has been observed in animal studies, but human information is not conclusive because of the relative novelty of radiation poison compared to the human life span.

Genetic mutations, stillbirths, and malformations of babies have been observed from humans exposed to radiation. Much data was obtained from the populace of the cities of Hiroshima and Nagasaki in Japan since 1945. Human sterility can result from a single high-dose exposure of intense radiation; long-term effects include an acceleration of the aging of reproductive structures. Cataracts develop in the lens of the eye just a few years after exposure to sufficient doses of radiation. At higher doses of radiation, lens damage occurs even more quickly.

Sources of Ionizing Radiation

Natural Background

There is a natural flux of radiation in our environment coming down from space (both cosmic rays and rays from our sun) and up from the earth. Fortunately, the earth's magnetic shield diverted much of this radiation; this, and the filtering effects of the atmosphere, allowed for the development of life on earth. The earth, itself, shields its surface from much of the subterranean radioactive events.

Humans, and all life, are continuously exposed to the low levels of radiation present in the environment. Typical sources are the materials in soils and rocks, as well as building materials. Materials in the environment are also inhaled, or ingested as food and water. Traveling to the mountains, living in Denver, or flying in an aircraft increases exposures to cosmic rays and, therefore, cancer or mutation risks. Cigarette smokers also are at increased risk because they have a dose of up to three times higher than natural background because of inhalation of the by-products $_{210}$Po and $_{211}$Po, which have accumulated in the tobacco.

Health Science Application

X rays, radiopharmaceuticals, and radiographic examination are widely used in the medical field. These applications are now closely regulated.

Nuclear Reactions

Enrico Fermi and his associates developed a process known as nuclear fission and created the first nuclear reactor in Chicago in 1942. In reactors, a *chain reaction* occurs in which one action initiates several others, which activate several others, and the activity increases exponentially with the subsequent release of a tremendous amount of energy. Nuclear energy can also be released by combining smaller nuclei into larger nuclei, fusion, but as yet an efficient and controlled method for collecting this energy has not been devised. The only practical demonstrations of fusion energy release have been hydrogen bombs. In the laboratory, fusion reactions are created; however, energy production, or making more than you invest (break even), remains elusive.

Nuclear Weapons

The Manhattan Project and other nuclear weapon testing started in 1945. Most of the testing is now underground, specifically because a ban on open-air testing was agreed to in 1963 (or abandoned entirely by some countries because of a current moratorium on testing). Risk of radiation exposure to humans from this testing is generally from a number of radioactive nuclides of Sr, Zr, Ru, I, Cs, Ce, and Pu, assuming you are not within the blast zone. Risk of contamination exposure is from the fallout of these atoms, from gamma rays, from assimilation of radioactivity through the food chain, and from beta doses to the skin.

Nuclear Power Production

Electric power is produced by harnessing the energy of a nuclear reaction. Use will likely increase over the next century; however, technology can probably never be developed to make the risk of catastrophic events less likely. Risks also stem from uranium mining, fuel processing, the reactor itself, and storage of the nuclear wastes. Nuclear weapons and power have left a hideous legacy that will always continue to plague humans. We say this because the masses of nuclear products we have already created will be with us for millions of years. Elements like plutonium are both highly radioactive and extremely toxic.

Other Sources

Television sets, luminous-dial watches, security systems, smoke detectors, electron microscopes, and building materials are among many consumer products emitting ions or containing radioactive materials. Air travel increases risk as cosmic radiation increases with altitude. Radon (Rn222) has become of concern recently as high levels have been found in building materials, in foundations, in the soil and rock underlying structures, and in water supplies to the home. These sources of ionizing radiation clearly have many negative effects on human health and safety, among them, lung cancer.

Doses From Natural Background Radiation

Natural background radiation comes from cosmic radiation of outer space that reaches the earth, radioactivity of the earth's crust, and inhaled or ingested radionuclides in our bodies. The cosmic radiation is made up of charged particles from our sun and tends to be attracted to the earth's poles. Thus, cosmic radiation is deflected from the equator to the north and south poles. The aurora borealis results from the movement of these charged particles down the lines of the earth's magnetic field. The level of radioactivity from the earth's crust varies greatly between areas. Where rocks have more uranium and thorium, humans will be exposed to more gamma radiation and thus experience a higher background radiation. Higher background areas are observed in Colorado, and lower background areas are observed in New York. Very small amounts of radioisotopes are inhaled or ingested and contribute to the background radiation our bodies naturally encounter. The radioactive isotope potassium–40 does make an appreciable contribution to our exposure; however, other radioisotopes are usually found in very low amounts. In the United States natural background levels vary from a high in Colorado of 220 mrem (millirem) to New York City, with 68 mrem. Worldwide background levels vary also with some high natural background areas being Brazil (coastal areas) 500 mrem and India (Madras State) 1,300 mrem.

Specific Radionuclides

Because the toxicity of radionuclides is very agent-specific, we provide only a brief overview.

Alkaline Earth Elements

The two alkaline earth elements of most interest are radium (Ra226), and strontium (Sr38). Radium is deposited in the bones, induces osteosarcomas of the skeleton, and carcinomas of the mastoid and paranasal air sinuses. Radium is found in the foundations of homes, earth and soil, water supplies, and certain building materials (particle board). Nuclear power generation accidents and weapons testing are the major environmental sources of strontium. There is a strong tendency for strontium to bioaccumulate in the food chain, where fallout from the atmosphere is picked up by vegetation (grass), which is eaten by sheep or cows. Humans drinking contaminated cow's milk incorporate the strontium into their bones, where it is in close proximity to the hemopoietic (blood-forming) tissue. Very low oral doses of strontium result in neoplasia of bone and related tissues.

Iodine

Radioiodine ($_{131}$I) is immediately absorbed into the bloodstream and concentrated in the thyroid. It can also lead to hyperplasia, adenomas, and thyroid carcinomas.

Lanthanides

The lanthanides are produced abundantly in nuclear reactors and in detonation of nuclear weapons. Inhalation results in lung cancer and liver and skeleton carcinomas.

Actinides

All of the actinides are radioactive, but those of most interest are $_{238}$U, $_{235}$U, $_{234}$U, and plutonium. In laboratory animals, death has been from pulmonary fibrosis and pneumonitis. The most common neoplasms are pulmonary-bronchial.

Radon

Radon is a daughter product of radium. As a gas it diffuses from the ground and accumulates in houses and enclosed spaces. While the exact risk of radon in the home is still being debated, it is agreed that radon is unhealthy, and sites with high radon levels should be mitigated or abandoned. Tight buildings aggravate the radon risk. Mitigation measures include increased ventilation, entrance barriers, and bypass flow paths.

Tritium

Tritium is a hydrogen isotope formed in the upper atmosphere by interaction of gases with cosmic rays; tritium is also prevalent in nuclear reactor effluents. Found in water supplies, as tritiated water, tritium is immediately absorbed by the skin, gastrointestinal tract, and lungs, and it distributes throughout the body as easily as water, thus producing body effects comparable to those of whole-body irradiation.

Radiological Protection

We are constantly being exposed to radiation. It comes from outer space (cosmic rays), natural radioactive decay in rocks in the earth's crust, and even from some smoke alarms (very low levels of radiation). When you go to the dentist and have your teeth x-rayed to find a cavity, you are exposed to ionizing radiation. We cannot avoid radiation, but we can avoid unnecessary exposure to ionizing radiation.

Humans can tolerate some exposure to radiation. As mentioned before, exposure beyond a "safe level" can cause dangerous and life-threatening effects. What is a safe level for ionizing radiation, and who said it was safe? Specific radiation protection standards are set by the National Council on Radiation Protection and Measurement (NCRP). The NCRP is a nonprofit corporation chartered by the United States Congress. Many federal government agencies, like the Environmental Protection Agency

(EPA) and the Food and Drug Administration (FDA), follow the recommendations for safe levels set by the NCRP or set levels even lower and safer for environmental workers. A worldwide radiation protection body is the International Commission on Radiological Protection (ICRP) that is independent of government affiliation that recommends procedures for handling radioactive materials safely and sets safe levels for exposure. In Publication 26, the ICRP recommended an annual dose equivalent limit for uniform irradiation of the whole body as 50m Sv (5 rem) for radiation workers and 5m Sv (0.5 rem) for individuals in the general public.

Federal government regulation of environmental releases of radioactive materials is based primarily on the Atomic Energy Act of 1954, the Energy Reorganization Act of 1974, and the Clean Air Act Amendments of 1977. The Atomic Energy Act of 1954 created a framework for federal control of civilian industrial use of radioactive materials. Private industry was regulated by the U.S. Atomic Energy Commission (AEC). The Energy Reorganization Act of 1974 abolished the AEC. The principal regulatory elements of the AEC were reestablished as the U.S. Nuclear Regulatory Commission (NRC), and the remaining elements were established as the Energy Research and Development Administration (ERDA), which was later incorporated into the U.S. Department of Energy (DOE). The Clean Air Act Amendments of 1977 gave the Environmental Protection Agency (EPA) the authority to establish air quality criteria and to define control techniques.

The Nuclear Regulatory Commission (NRC) regulates the nuclear industry pursuant to the Atomic Energy Act. The NRC regulations are included in "Title 10 of the Code of Federal Regulations" (10 CFR) and have the force of law. In addition, the NRC publishes "Regulatory Guides" that describe acceptable methods for compliance with specific regulations.

The Department of Defense (DOD) regulates most of its own activities involving nuclear weapons and nuclear-powered vessels. However, most other activities concerning radioactive materials are regulated by the NRC. The Department of Energy (DOE) facilities and contractor-operated national laboratories are not subject to NRC regulations, but they follow the DOE manual of nuclear regulations, which is consistent with NRC regulations and EPA guidelines.

In an effort to reduce the health risk to environmental workers, the concept of ALARA (As Low As Reasonably Achievable) is used to state the philosophy of protection. Environmental workers may, at times, be required to wear respirators, disposable coveralls, disposable shoe covers, disposable rubber gloves, and eye protection, as well as to provide a urine specimen. Thus, health professionals can test for internal radiation contamination. These actions are all done with the workers' protection, health, and safety in mind. By reducing the radiation exposure to skin, lungs, and eyes, the risk factors to the worker are lessened.

Conclusion

Radiation is ever present in our environment. As workers clean up the environment, they may encounter radioactive waste or mixed waste (radioactive and nonradioactive together).

The workers must protect themselves with respirators and other protective clothing. The purpose of this protective clothing is to prevent internal exposure of radioactive isotopes within the body. The workers must be careful of inhalation through the lung, ingestion (no eating near the clean-up site), or absorption through the skin. The internally deposited isotopes are very dangerous because they produce continuous radiation exposures to cells until the isotope undergoes radioactive decay or is excreted by the body. Because continuous radiation exposure to cells can cause health problems, exposure in workers must be controlled. Laws are in place to limit worker exposure, lower the hazards created by the energy of the emitted radiation, and decrease the harmful biological effects of the radiation.

SUMMARY

▸ Radioactive materials are commonly used in industry, medicines, and power plants.
▸ There are several measurements of radiation, including roentgens, curies, becquerels, rads, grays, rems, and sieverts.
▸ Some of the biological effects of ionizing radiation are acute radiation syndrome and cell damage and death.
▸ To provide radiological protection, the U.S. government has established several groups to set standards for the safe use of radiation and to monitor radiation levels in the workplace and the environment. Some of these organizations include the National Council on Radiation Protection and Measurement (NCRP), the Environmental Protection Agency (EPA), and the Food and Drug Administration (FDA).
▸ Other radiologic protection groups include the ICRP, the NRC, the DOD, and the DOE.

STUDY QUESTIONS

1. What is a roentgen (R)?
2. What is a curie?
3. What is a becquerel?
4. What is a rad and gray?
5. What is a rem and sievert?
6. List and describe the four stages of acute radiation syndrome.
7. From what groups of people did we gather information about the chronic effects of ionizing radiation?
8. What are the three sources of natural background radiation for humans?
9. What is the role of the NCRP?
10. What is the ALARA concept?

10

Common Industrial Hazards

Douglas Stutz

COMPETENCY STATEMENTS

This chapter describes some common industrial hazards such as blast injuries, irritant and toxic gases, methemoglobin producers, corrosives, solvents, and toxic metals. When readers have finished studying this chapter, they should be able to meet the following objectives:

▶ Describe the effects on the human body of blast wave injuries.

▶ Identify some irritant gases and discuss their toxic effects on the body.

▶ Identify some toxic gases and discuss their toxic effects on the body.

▶ Describe the differences between the effects of irritant and toxic gases.

▶ Identify some of the producers of methemoglobin and describe their effects on the body.

▶ Describe the health hazards posed by corrosives.

▶ Describe the health risks of exposure to solvents.

▶ Describe how people can be exposed to harmful levels of metals and identify some harmful metals.

Portions are reprinted with permission from HAZTOX, EMS Response to Hazardous Materials Incidents, by D. R. Stutz, Ph.D. and S. Ulin, MD, GDS Communications, Miramar, Florida, 1994.

INTRODUCTION

While health effects, toxins, and chemical hazards are the primary subject of this text, there are specific hazards worth discussing individually. These chemicals, after acute events, can cause CNS effects ranging from headache to coma and convulsions; lung effects ranging from bronchospasm to pneumonia; cardiovascular effects including arrhythmias and shock; and gastrointestinal problems ranging from nausea to hepatitis. Other problems resulting from absorption include renal failure, dermatitis, burns of the skin, corneal burns, and hemolytic anemia.

BLAST INJURIES

Major health hazards fall into a wide variety of areas. Consider, for example, the explosive capability of most chemical and hazardous agents. One must consider that the blast effects will probably include a vast amount of conventional injury or trauma. One of the effects of a blast is a blast wave. A brief but violent blow strikes the entire surface of the body on the side of the shock wave. Much of the energy is reflected; however, part of it is transmitted through the tissues and strikes the internal organs, which vary in their susceptibility to such a blow.

A *blast wave* is an overpressure that occurs as a result of the very rapid buildup of pressure. The overpressure passes over and through the body very quickly, in fact in an almost incomprehensibly fast time frame. The wave moves out from the point of the explosion in a bubble shape (see Figure 10.1). As it moves, not only is an overpressure created in front of the wave, but a vacuum occurs behind it. This action causes the tissues and the organs to vibrate and move violently, thus causing injury. The farther away one is from the source of the blast, the safer one becomes. As an example, Table 10.1 provides data on the overpressure created by an explosion caused by 30 kilograms (66 pounds) of TNT and one caused by 50 kilograms (110 pounds). As you can see, distance is critical in determining the amount of damage done.

The solid tissues are virtually incompressible and simply vibrate as a whole, thus escaping serious injury. Organs containing gas, such as the middle ear, lungs, and gastrointestinal tract, are liable to damage because they are compressible and have tissue-gas interfaces.

Compressibility means displacement, and wherever tissues of differing densities lie side by side, this displacement may cause distortion and tearing. Lesions are most severe at junctions between tissues and at sites where loose, poorly supported tissue

▶ TABLE 10–1
Blast wave pressure.

MAXIMUM PRESSURE OF BLAST WAVE		
Charge	4 Meters (4.4 yards)	15 Meters (16.4 yards)
30 kg (66 lb)	100 psi	6 psi
50 kg (110 lb)	200 psi	10 psi

▶ **FIGURE 10–1**
Blast wave configuration.

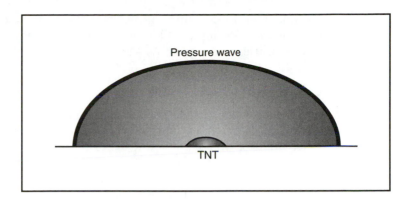

is attached to dense tissue and is displaced beyond its elastic limits. Clinical experience and animal experimentation have shown the most vulnerable organs to be the eardrums, lungs, and gastrointestinal tract.

Injuries Caused by Blast Effects

Injuries can be caused in virtually every body system when someone is subjected to a blast wave. Table 10.2 provides the average overpressure required to initiate certain injuries. As you can see, not a great deal of pressure is required to cause damage. The systems of the body are not equally susceptible to the blast wave. The major systems are listed in the following sections in decreasing susceptibility:

Auditory System

In the auditory system damage can occur in three ways:

1. Rupture of tympanic membrane (7 psi in adults)
2. Dislocation of ossicles
3. Damage to the inner ear

As an example, in a Philadelphia refinery, a fire broke out in the still used to manufacture isopropylbenzene. The leak was caused when a pipe ruptured and allowed escaping flammable liquid to vaporize. The gas was ignited by an open flame about 45.7 meters (150 feet) away and flashed back to the still. The fire ruptured a steam

▶ **TABLE 10–2**
Short duration pressure effects on unprotected persons.

Pressure (psi)	Effect
5	Slight chance of eardrum rupture
15	50% chance of eardrum rupture
30–40	Slight chance of lung damage
80	50% severe lung damage
100	Slight chance of death
150–180	50 % chance of death
200–250	Death usually

line under 600 psi pressure. The noise created by the escaping steam made communications virtually impossible. The only way to get any orders through was to send a runner or use hand signals. As a result of this noise, 11 firefighters were treated for acoustic trauma.

Lung Tissue

In the lungs hemorrhages usually occur on the side turned toward the explosion. It varies from pinpoint hemorrhages in survivors to massive intrapulmonary hemorrhages in those who die.

Solid Viscera

Injuries to the solid viscera are usually caused by the violent acceleration and deceleration forces of the blast rather than by the pressure wave.

Gastrointestinal System

Damage to the gastrointestinal system is unusual in a surface blast and is more common underwater. Closed injuries usually take the form of perforation of the air-containing viscera, large bowel, stomach, or multiple hemorrhages in the walls and lumen.

Nervous System

Experimental evidence strongly supports that CNS changes are caused by cerebral air embolism, which originates from pulmonary damage.

Coronary Artery Air Embolism

Arterial air embolism has been shown to be one of the major causes for sudden death as a result of a blast.

Traumatic Amputation

In explosions in confined spaces, it is common for 25% of the dead or severely injured to have lost all or part of a limb.

IRRITANT GASES

A specific "toxic inhalation" may produce one, some, or all of the following processes. Vapors (inhalation of a toxic substance) liberated during an ammonia release from a storage tank leak may produce only a pulmonary irritant with no significant asphyxia, systemic toxicity, or thermal burn. Smoke liberated during an apartment house fire, however, does most of its damage through carbon monoxide (a chemical asphyxiant), hydrochloric acid (a pulmonary irritant elaborated from the heating of polyvinyl chloride), and organic acids and aldehydes adherent to carbon soot particles (another pulmonary irritant). Irritant gases cause symptoms to the eyes, nose, the throat and can cause watering, burning, and sometimes pain. In some cases, because these are caustic gases, they can cause burns and corneal abrasions. The gases that are water-soluble are much more likely to cause upper respiratory effects and less likely to cause lower respiratory effects unless in a massive exposure.

The upper respiratory effects are predominantly cough and bronchospasm. These may manifest clinically by wheezing and pathologically by tracheal/bronchial inflammation. If the irritant gas is of significant concentration and exceeds the scrubber mechanism of the upper respiratory tree, the patient may develop lower respiratory symptomatology. Additionally, if the gas is not particularly soluble in water, it may escape the scrubber mechanism of the upper respiratory tree and be more likely to cause lower respiratory problems, primarily pulmonary edema. Lastly, some of these gases have been known to cause hypersensitivity reactions on subsequent re-exposure.

Decreased inspired O_2 content in inhaled gas leads to simple asphyxia. Where there is no oxygen to breathe, there is no life. Oxygen may be used up in combustion or displaced by another gas. The mechanism of death by asphyxia is probably through secondary circulatory arrest that speeds the brain death.

The high temperature of inspired gas leads to thermal burn of the respiratory tree. The presence of pulmonary irritants leads to chemical burns of the respiratory tree. The presence of cellular toxins leads to chemical asphyxia. The respiratory tree has a limited number of sites that can be harmed by thermal exposure or chemical irritants and a limited number of ways of manifesting damage. The respiratory system can be divided into four parts: upper airway, trachea and mainstem bronchi, smaller bronchi and bronchioles, and alveoli. The upper airway responds with swelling that results in hoarseness, sensation of choking and constriction, and eventually full obstruction. The trachea and major bronchi respond with swelling and outpouring of secretions that result in coughing and chest pain. Subglottic problems, such as severe tracheitis, can appear as upper airway obstruction. The smaller airways and bronchioles respond with swelling, spasm, and outpouring of secretions and mucous plugging that result in wheezing. The alveoli respond by developing a leaky membrane with subsequent alveolar edema that results in tachypnea, dyspnea, and cyanosis. In general, upper airway problems occur earlier in the course of a "toxic exposure" (exposure to a toxic substance, subsequent absorption, with a dose of the substance taken into the body) than does alveolar edema, which may be delayed up to 48 hours after exposure.

Thermal damage, in general, is localized to the upper airway. Heat is dissipated rapidly as gas passes down through the nose, mouth, larynx, pharynx, and trachea. Gas at 500°C (932°F) at the vocal cords becomes 50°C (122°F) at the tracheal bifurcation. Wet heat penetrates further down the respiratory tree than does dry heat; so always carefully evaluate a superheated steam exposure for laryngeal and tracheal damage.

The most common toxic gases that act as pulmonary irritants are chlorine (Cl_2), phosgene ($COCl_2$), ammonia (NH_3), sulfur dioxide (SO_2), and nitrogen dioxide (NO_2). In addition to their intrinsic toxicity as gases, as they enter the tracheobronchial tree, they combine with water and become converted to, e.g., NH_3, and ammonia to ammonium hydroxide (NH_4OH). Sulfur dioxide becomes sulfurous acid, and nitrogen dioxide becomes nitric acid. Use of a water-soaked towel over the eyes and mouth and nose by rescue personnel is logical in that it permits conversion of the gases to their caustic counterparts outside of the body's mucous membranes and endothelium.

The severity of the toxic effects due to exposure depends on absorptive factors such as these:

1. Solubility of gas in water. The more soluble the gas (such as ammonia), the more that the mucous membranes of the mouth, conjunctiva, nose, and the upper airway are irritated. The less soluble the gas, the more that the lower airway and lungs are affected (for example, phosgene). Certain gases of intermediate solubility (such as chlorine) affect all parts of the respiratory tree. The soluble gases that affect the upper airway and mucous membranes give warning of their presence by the intense eye, nose, and throat burning they produce. Insoluble gases like phosgene cause significant pulmonary damage without producing any warning symptoms of their presence.
2. Duration of exposure
3. Concentration of gas
4. Presence of underlying pulmonary disease

The effects of a few common gases include the following:

1. *Hypoxia:* Butane, CO_2, methane, natural gas, cyanide
2. *Poison:* Carbon disulfide, CO, methyl bromide, hydrogen cyanide
3. *Pulmonary irritants:* Ammonia, chlorine, sulfur dioxide, nitrous fumes
4. *Systemic effect and pulmonary irritant:* Acetylene, hydrogen sulfide, ozone, and chlorine

MAJOR GASES THAT CAN CAUSE INJURY

Acrolein

Acrolein is a by-product of combustion of fuels and particularly wood burning. You might see patients exposed to this from fires or fireplaces where the smoke is out of control. For example, in a case in which two children died from an overheated fryer, autopsy reports showed they had cellular desquamation of their upper respiratory tree and pulmonary infarcts. Their deaths may not have been caused by pure acrolein inhalation, as other products were undoubtedly present. Asthmatics and individuals with bronchospastic chronic obstructed pulmonary disease (COPD) are particularly susceptible to this gas.

Sulfur Dioxide

Sulfur dioxide is produced from oxidation of organic materials containing sulfur. Sulfur dioxide is a strong irritant that causes pulmonary edema, producing effects similar to those of cyanide. It is a by-product of smelting and paper manufacturing.

Sulfur dioxide is one of many gases that are important in air pollution. It reacts with the moist membranes of the upper respiratory tree to form an acid. Additionally, it apparently acts through a vagal (tenth cranial nerve that supplies thoracic and abdominal viscera) mechanism and can result in significant bradycardia as well as causing smooth muscle contraction. Asthmatics who have hyperreactive airways might be more sensitive. Studies have shown that patients exposed to significant sulfur dioxide will suffer a chronic, obstructive defect. This is detected in serial pulmonary function studies. Bronchial hyperreactivity is also observed. Sulfur dioxide levels are one of the markers used in times of temperature inversions and as a warning to patients who have reactive airway disease, who can be warned to remain in a more comfortable environment.

Hydrogen Chloride

Hydrogen chloride is released through thermal degradation of many products; however, polyvinyl chloride is one of the most common and is of concern to firefighters. The effects are similar to those of chlorine gas. It is absorbed in soot, and the hazard depends partly on particle size. It can produce respiratory distress either early or late. Preventricular contractions of the heart often occur as a result of the exposure. Even though these effects can be a result of inhaling the gas, it is primarily classified as an irritant.

Ammonium Nitrate

Ammonium nitrate is used in dynamite and blasting agents as well as fertilizer manufacture. In gaseous form, it is colorless and extremely toxic. The fertilizer grade decomposes into nitrous oxide and water; this reaction liberates a great deal of heat, which increases decomposition and may cause an explosion. The symptoms of exposure are tremor, slurred speech, ataxia, stupor, and coma. Liver failure may be the clinical result.

Ammonia

Ammonia is a common irritant gas. It is used as a fertilizer, both in the gaseous form and combined with other compounds. It is also used as a refrigerant and in cleaners. Its use as a refrigerant will become much more common as the use of freon as a coolant decreases. Ammonia is a gas at room temperature. It is transported or stored compressed into a liquid and dehydrated, which accounts for the term *anhydrous,* which is most often associated with it. Reaction with water forms a product with a pH of 9.25, which is considerably caustic. It causes thermal damage by burning or frostbite as it removes heat from whatever it touches. Additional damage is caused by its caustic effects. The nose will retain a significant amount of ammonia. Once in the body, the gas is absorbed and converted to urea. Mental status changes may occur within a short period as a result of accumulation of urea. These changes usually last only 30 minutes to one hour.

Ammonia is a soluble, colorless gas; is an alkali with a pungent odor; is extremely irritating to the eyes, mucous membranes, and respiratory tract; and is often used with refrigeration equipment. Inhalation of concentrated ammonia produces asphyxiation in minutes; lesser amounts produce pulmonary edema and chemical bronchitis. Anhydrous ammonia (NH_3) is a colorless, alkaline, nonflammable compressed gas with a sharp, pungent odor. It is normally stored in two-layer storage vessels. Upon release from a pressure vessel, liquid ammonia vaporizes and expands rapidly, absorbing heat with the expansion. It irritates skin and mucous membranes.

The primary symptoms upon inhalation are burns to the upper airway (sometimes leading to obstruction), blurred vision, corneal abrasions, pulmonary artery thrombosis, muscle spasms and weakness, hemorrhagic necrosis, and change in mental status. All of these have been reported from inhalation exposures. In fact, a death has been reported from an exposure to the eye. Patients who are dosed orally might have burns of the GI tract, acidosis, and renal failure along with burns to the skin, corneal damage, and even death.

Methyls (CH$_3$)

Methyls are used as refrigerants and in organic synthesis and fumigants. Methyl bromide is used as an herbicide, and can penetrate rubber gloves. Methyl chloride is used as bubbling fluid in ornaments. These gases are tasteless and odorless and are CNS depressants similar to chloroform. They can cause liver, renal, and myocardial damage. Symptoms produced by these agents include fatigue, headache, dizziness, ataxia, and weakness. Early deaths are usually secondary to pulmonary effects, in a way similar to chlorine; late deaths are usually secondary to renal failure.

Chlorine

Chlorine is a greenish-yellow gas that has been responsible for many of the deaths and injuries that occur because of hazardous materials. It is one of the more common irritant gases and is found around the home, around swimming pools, and in the manufacture of various products. It is detectable in higher concentrations and, when chlorine combines with moist membranes of the upper respiratory tree, it forms acids. When bleach, which is a mild form of a chlorine acid, is mixed with a stronger acid, chlorine gas can result. That chlorine gas can react with moist upper respiratory membranes and cause irritation. More commonly, when bleach is mixed with ammonia in the household setting for cleaning, the two can liberate a chloramine gas that reverts in the distal airways to hydrochlorous acid and ammonia, both of which can be very irritating and can cause not only upper, but also lower, respiratory effects. Additionally, the reactions can result in radical formation or loss of surfactant. Free radical formation may occur. This gas rarely causes chronic effects, and patients are usually asymptomatic in a short period of time.

Chlorine is a nonflammable compressed gas used as an agent for bleaching cloth and in paper and water purification. It may be produced as a by-product of a chemical reaction such as mixing household cleaners (such as Clorox) with acids (vinegar) or mixing bleaches with household ammonia. Chlorine is heavier than air, acts rapidly, and becomes corrosive when mixed with water, such as in the respiratory tree. Death is usually secondary to sloughing of bronchial epithelium and obstruction of the airway. A high concentration will result in chemical burns of the skin and mucous membranes.

Pre-hospital treatment of chlorine exposure includes the following steps:

1. Self-protection, such as protective clothing (running gear and SCBA equipment, for example)
2. Decontamination
3. Humidification of oxygen
4. IV therapy
5. Bronchodilatation
6. Eye irrigation with normal saline
7. Copious skin irrigation with normal saline
8. Pulmonary edema (per protocol)
9. Other supportive care as necessary

Formaldehyde

Formaldehyde is a contaminant of smog, but is most commonly found in the manufacture of foam and is significant in foam insulation. There is particular concern in its use in trailer homes and new construction. It is also used as a disinfectant and fixative. It is a colorless gas, but definitely has olfactory warning properties. In the environment, it accounts for 50% of the total aldehyde exposure. When it is mixed with methanol to form formalin, the total methanol is approximately 15%. This is of particular importance if you are dealing with a patient who had ingested formalin. A methanol level should be obtained. It has been debated whether formaldehyde is an allergen. Some people appear to become or be sensitive to it. Most authorities agree that it can cause asthma and squamous cell cancer of the nose.

Ozone

Ozone is a common component of our environment; however, as a pollutant in the air, it becomes a major problem. Ozone causes environmental effects such as smog, and it increases the sensitivity of the body to histamine, acetylcholine, and allergies. It increases the fragility of the red blood corpuscles, causing mild hemolysis. Ozone causes irritation of the upper airways; however, it is considered to be a deep lung irritant, where it is a significant cause of pulmonary edema.

Because ozone is produced in so many ways, it is easy to miss it as a cause of a toxic reaction. For example, ozone is produced in ways that are as innocuous as using a copy machine; ozone generators are even being used in swimming pools, for hot tubs, and in the home as a means of biological control. Rescuers must be aware of the possibility of ozone exposure beyond that received through air pollution.

Nitrogen Dioxide

Nitrogen dioxide is a product from fires (particularly mattresses) and is a reddish-brown gas. Interestingly, it is also generated in closed environments where crops high in nitrates are stored, such as silos. Silo fillers disease, first described in the 1950s, occurs predominantly in the fall when crops are ground up and put into the silo. The greatest risk usually occurs in the first several hours after filling a silo, but was delayed as long as six weeks in one particular case. The nitrogen dioxide produces acids on contact with water and, for that reason, may cause skin burns and metabolic acidosis. Because the nitrogen dioxide is decomposed into nitrates that are fermented into nitrites, the chemical may cause methemoglobinemia and vasodilation. Significant exposure may result in a pulmonary edema. As much as a day may elapse before the development of edema.

Isocyanates (MIC, TDI)

The isocyanates, such as methyl isocyanate, which caused so much death in Bhopal, India, are used in the manufacture of plastics and pesticides. They can also be significant by-products of thermal degradation of these products during fires. Despite its name, cyanide is not released from the isocyanates. The effects are primarily respiratory; however, dermal effects can occur in the form of contact dermatitis. Blistering of the skin is possible. Respiratory effects include chemical bronchitis, asthma caused by the isocyanate, acute nonspecific airway disease, and chronic nonspecific airway disease. A hypersensitivity pneumonitis and neurological effects, which are manifested by a drunken appearance, numbness, or loss of balance, may also occur. Some of these symptoms may exist for an extended period of time.

Hydrogen Fluoride

Hydrogen fluoride is a colorless gas used in a variety of manufacturing processes and around the home to remove rust or clean different metals. It is caustic, causes coagulation necrosis, and complexes with calcium to cause local or systemic hypocalcemia (low calcium levels), which results in the release of potassium and subsequent pain. Animal studies have shown that the nose will basically scrub out the majority of the hydrogen fluoride as long as there is no significant hyperventilation. If there is less nasal breathing and significant hyperventilation, more upper airway effects will be seen. Patients may develop ventricular arrhythmias due to the hypocalcemia. Pulmonary edema has been also reported. It is important with this particular gas to recognize potential hypocalcemia and treat patients with calcium gluconate. (Calcium gluconate is less irritating than calcium chloride.) Although they may have minimal symptoms, patients who have been exposed by way of chronic inhalation may have skeletal abnormalities from the leaching of the calcium. (See discussion on hydrofluoric acid, under "Corrosives.")

Fluorocarbons

Fluorocarbons, such as 1,1,2-trichloro-1,2,3-trifluoroethane (freon) can be described as irritants as well as toxic gases. They are commonly used in refrigerants. As propellants, they were banned in nonessential products in 1979. These compounds often exist as gases at room temperature and are often compressed to become liquids. As such, when they expand, they attract heat from their surroundings and can cause frostbite and airway edema. These compounds are primarily complexes of fluorine and chlorine and can degrade into chloride, hydrofluoric acid, or phosgene. As such, they can cause many of the same symptoms discussed for those gases. The fluoroalkenes (such as tetrafluoroisobutylene) are primarily lung irritants. However, the majority of fluorocarbons are fluoroalkanes that rarely cause pulmonary toxicity. They can cause a systemic toxicity by sensitizing the heart to catecholamines, resulting in escape rhythms and ventricular fibrillation. Patients may develop heart-slowing and systole, perhaps as a vagal response. Additionally, fluorocarbons can cause asphyxia in patients who are huffing or bagging these vapors to get high. Unique to the irritant gases in this case is the concern that one must be cautious about the use of adrenergics in treating these patients because of the sensitive myocardium. Atropine is often considered ineffective; however, it has been used with success.

TOXIC GASES

In contrast to the irritant gases that cause their damage primarily to the lungs and have significant immediate symptomatology, toxic gases may cause pulmonary toxicity that is often delayed or that has systemic manifestations. Included in this differential are many pulmonary edema formers. Although many irritant gases can cause pulmonary edema, two that are most likely to do so are ozone and nitrogen oxides. Some solvents have been reported to cause pulmonary edema. Pulmonary edema is a common finding in humans who have been exposed to high levels of inhaled carbon tetrachloride. The effects are often delayed and may be related to renal dysfunction. Methylene chloride may produce pulmonary edema secondary to phosgene generation. Some of the ethers may cause pulmonary edema and renal dysfunction. Trichlorethylene has produced pulmonary edema in an industrial worker, and laundry workers have had pulmonary edema when exposed to tetrachloroethylene.

Simple Asphyxiants

Simple asphyxiants are chemically inert gases that occur either naturally or as breakdown products. For example, nitrogen is found in mining; carbon dioxide is a fermentation product found in mine shafts and dry ice storage; and methane is found in mine tunnels and enclosed organic decomposition (for example, in liquid manure storage tanks). Natural gas is made of methane and ethane, whereas bottled gas is made of butane and propane. Simple asphyxiants effectively reduce the O_2 of the inspiratory air and through the mechanism cause hypoxia. Seventeen percent inhaled oxygen is considered to be the safe limit for prolonged exposure; however, tolerance depends on the patient's total oxygen-carrying capacity. Ten percent O_2 will often cause symptoms of dizziness, dyspnea, and tachypnea, and 7% stupor (5% stupor is considered life-threatening).

Carbon Monoxide

Patients who are inadequately treated for carbon monoxide poisoning (approximately 1% of the severe carbon monoxide exposures) may have delayed neuropsychiatric effects, including aphasia, akinesia, apathy, disorientation, hallucinations, Parkinson-like symptoms, and incontinence. These patients may even show up in emergency rooms or in primary care facilities after a lucid interval of as long as 21 days. This long period of onset suggests the importance of following patients on a weekly basis after their exposure. EMS responders must recognize the possibility of delayed symptoms. The incidence of this condition varies in the literature. It is reported at somewhere between 3% and 40%.

Cyanide

Cyanide is a product of nitrogen combustion and is an intermediate in the production of nylon. It is a by-product during the synthesis of acrylonitriles. Cyanide is used as an insecticide and rodenticide for fumigating in enclosed spaces. Cyanide salts are used in electroplating and metal treatment. The major source of cyanide released to the air is vehicle exhaust. Fires involving polyurethanes, silk, and wool will possibly produce cyanide. Additionally, cyanide gas may be released when an acid is added to a cyanide salt. As a gas, it is rapidly absorbed via inhalation. As a gas, it can exist in three forms: hydrogen cyanide, cyanogen, and cyanogen chloride. A common occurrence is when acid and cyanide solutions are accidentally mixed in electroplating baths and large volumes of HCN are released. Cyanide exists primarily as HCN with the body at physiological pH.

Cyanide along with carbon monoxide and methemoglobinemia reduce the potential delivery of oxygen to the tissues, but through different mechanisms. Because of these different mechanisms, there may be a decreased central arterial venous oxygen saturation gap that may be one of the few laboratory clues available to detect cyanide poisoning. Generally, signs and symptoms are nonspecific and resemble those of anxiety or hyperventilation syndrome. There can be considerable GI effects, pulmonary edema, EKG changes, coma, and seizures. Cyanide will combine with a number of human enzymes; however, its greatest toxicity seems to be caused by its great affinity for ferric iron in cytochrome oxidase. This enzyme is involved in cellular adenosine triphosphate (ATP) production. ATP is the major energy source in living cells and is produced in the mitochondria of the cell through an electrotransfer process known as oxidative phosphorylation. The enzyme cytochrome oxidase has two subcomponents, cytochrome a and cytochrome a_3. Hydrogen cyanide will combine with a_3 where it interrupts electron transfer to the oxygen molecule. As a result, oxygen consumption and oxidative phosphorylation are significantly decreased. ATP must then be produced through other methods within the body. Usually this is an anaerobic (without oxygen) process. Metabolic acidosis increases. The cytochrome oxidase in the heart cells is more sensitive to inhibition by cyanide than other cells within the body. The cells in the body most sensitive to energy reduction—the brain and the ear—are the first to show the effects of cyanide.

The body is able to remove cyanide (using a Phase II reaction) by transferring sulfane sulfur to cyanide, which makes it thiocyanate. This is then removed from the body through the kidneys and passed in the urine. Several methods have been theorized about how cyanide is removed. It is believed that two enzymes, rhodanese and B-mercaptopyruvate-cyanide sulfurtransferase, assist in transferring sulfur from thiosulfate to cyanide and mercaptopyruvate respectively. Other work suggests that there are a number of sulfurtransferases, including those above, that form sulfane sulfur, which will form complexes within the albumin in the blood. It is believed that this complex will then form thiocyanate. Cyanide also has a significant affinity for the cobalt ion (Co^{2+}) and will combine with hydroxocobalamin to form cyanocobalamin. This complex is then excreted through the bile and urine. The use of large quantities of hydroxocobalamin as an antidote is suggested in several countries, but not within the United States.

Cyanide produces cellular hypoxia producing symptoms including confusion, dizziness, and collapse. Little or no improvement will occur with the administration of oxygen. Amyl nitrite pearls (small round ampules) are used as initial treatment followed by administration of sodium nitrite IV. There is controversy about the efficacy of amyl nitrite pearls; however, if sodium nitrite IV is not available, they can be used during the IV setup period. Sodium thiosulfate is administered after the sodium nitrite.

Acrylonitrile

Acrylonitrile is an explosive, highly flammable liquid compound used to produce plastics and rubber and is found in many super glues. It is also an intermediate in pesticide manufacturing. Because of its volatility, it is an irritant affecting the eye, mucous membranes, and skin. During its metabolism, cyanide can be released. It can be absorbed by the skin, the lungs, and through ingestion and may penetrate rubber and leather gloves. Since it acts much the same way as cyanide (by creating cellular asphyxiation), an acute exposure should be managed with amyl nitrite or sodium thiosulfate. Some case reports of acute toxicity resemble cyanide exposure. The more common event, however, is from contact dermatitis or where the skin becomes glued together. It is a suspected carcinogen.

Hydrogen Sulfide

Hydrogen sulfide is a toxic gas commonly found in sewers and is the product of putrefaction, the shale oil industry, areas of volcanic formation, and the manufacture of fertilizer. Seventy different occupations have potential exposure, especially in the petroleum industry. It is highly toxic, with effects ranging from respiratory failure to death. The effects of low concentration are eye irritation, cough, shortness of breath, and pulmonary edema. An effect of high concentration is initial CNS stimulation followed by CNS depression, headache, nausea, and vomiting with the patient becoming emotionally labile. Doses, which are synonymous with exposure due to complete absorption of gas, cause cellular anoxia with effects similar to cyanide. Hydrogen sulfide (the sulfide itself) binds with the iron in cytochrome oxidase causing an inability of the cell to absorb and use oxygen, which actually stops cellular aerobic metabolism. The treatment is similar to that for cyanide.

Hydrogen sulfide is frequently associated with the smell of rotten eggs. However, because one may become accustomed to the smell, smell is not an adequate indicator of its presence or concentration. It is the leading cause of death in the workplace related to toxic inhalation. It has been known to cause death after agitation of underground liquid manure tanks and the addition of sulfuric acid to a drain.

Toxic effects depend on the concentration and duration of exposure. With exposure to levels exceeding 1,000 ppm, victims develop respiratory paralysis and hypoxia. Other signs and symptoms are strictly dose-related. Because of its ability, in high concentrations, to cause paralysis of olfactory sensors, the warning sign of the "rotten egg smell" may last only momentarily. For this reason, it is known to cause death both to the initial victim and later to the rescuer. (The threshold at which the odor is detected is generally considered to be 0.13 ppm.) Hydrogen sulfide has been responsible for the deaths of many sewer workers and miners. Because it is heavier than air (vapor density of 1.19), it tends to lie in lower areas.

The primary route into the body is respiration, but irritation of mucous membranes is also a common injury site. It is considered to be more potent than cyanide, at least in studies using purified preparations of cytochrome oxidase. Resulting symptoms include cough, dyspnea, hemoptysis, pulmonary edema, confusion, vertigo, hypoxia, seizures, and coma. Most deaths occur at the scene of exposure. Victims who reach the hospital with vital signs usually survive unless other complications such as hypoxic encephalopathy are present (see Table 10.3).

Arsine

Arsine (hydrogen arsenide, H_3As) results from the reaction of hydrogen with arsenic in the refining of nonferrous metals. A colorless, nonirritating, flammable gas, it has a faint garlic smell. It inhibits tissue oxygenation, causes hemolytic anemia, and produces secondary effects such as acute renal failure and pulmonary edema. Symptoms are usually delayed a few hours and are nonspecific or related to the hemolysis. Symptoms include abdominal pain, hematuria, jaundice, damage to the kidneys, lungs, and heart. Chelation of this heavy metal gas offers little relief; however, an exchange transfusion may be of benefit.

▶ TABLE 10–3
Physical effects of hydrogen
sulfide.

PHYSICAL EFFECTS OF HYDROGEN SULFIDE	
Concentration (ppm)	**Physical Effects**
0.025	Detectable odor
10	Obvious and unpleasant odor
50	Conjunctival irritation first noticed
100	Loss of smell in 3 to 5 minutes
150	Olfactory nerve paralysis
300–500	Imminent threat to life
700	Dizziness, breathing stops, artificial respiration needed
900+	May cause instant death
1000	Rapid collapse, nervous system paralysis, imminent death

Stibine

Stibine is a gas that is liberated when antimony is treated with hydrogen, such as in the application of an acid. It creates a hemolytic anemia that is essentially the same as arsine, but, in addition, may cause burns and pruritic papules known as "antimony spots" that progress to pustules. As mentioned with arsine, an exchange transfusion may be the treatment of choice.

Metal Fume Fever

Metal fume fever is an occupational hazard, especially for welders. It results from a number of different metal oxides, particularly if the oxide is fresh. As opposed to poisoning from a gas, this condition is caused from particulate matter and causes a hypersensitivity reaction. The classic presentation is the worker who develops tolerance during the week because of the repeated exposure, and then after a weekend away from the workplace is re-exposed on Monday morning and develops symptoms that evening. Symptoms then are usually delayed 4 to 8 hours after exposure and may last as long as two days. The patient typically will have a cough, dyspnea, chest pain, fever, or pulmonary edema that resembles a pneumococcal pneumonia. Early symptoms may include a metallic taste in the mouth, which helps in the diagnosis. Patients who see a physician based on flu symptoms should be queried about their possible exposure to metal oxides. Fortunately, the syndrome is self-limiting and the treatment is life supportive. To enhance productivity and reduce the number of sick days, the workplace should be evaluated and actions taken to eliminate the exposure.

Phosgene

Phosgene was first synthesized in 1812 and was used extensively during World War I where it caused perhaps as many as 80% of the deaths by gas. It is probably used because of its mild irritating effects and the fact that it is heavier than air and therefore sinks into the trenches. It is primarily used in the manufacture of isocyanates, dyes, and insecticides. Phosgene is created by heated organic compounds decomposing in the presence of chlorine. It is also a by-product of combustion of a number of chlorinated hydrocarbons including PVC plastics and carbon tetrachloride, which is no longer used in fire extinguishers. It is classified as a pulmonary irritant.

Phosgene is a colorless gas that has a musty odor resembling fresh-mown hay or green corn. It decomposes into carbon monoxide and hydrogen chloride. Because it is not very soluble in water, it causes minimal irritation in the upper respiratory tree. Therefore, most patients who have been exposed are asymptomatic and have a latent period that typically lasts a few hours, but may be as long as 24 hours. The initial symptoms are mild and transient and consist of burning eyes, cough, and shortness of breath. Delay in symptom presentation often results in the development of a severe pulmonary edema and hemolysis. Death occurs secondary to anoxia. Because patients develop a secondary infection with bronchitis and pneumonia, prophylactic antibiotics have been recommended. N-acetylcysteine (NAC or Mucomyst) may act as a sulfur donor to trap the phosgene and convert it to a less harmful metabolite. Steroids have been used and are felt to be of value.

Phosphine

Phosphine is used in the manufacture of silicone crystals and the production of acetylene gas. Phosphorus in solid form may be used as a grain preservative or rodenticide. Aluminum phosphide fumigants and zinc phosphide rodenticides release phosphine gas upon contact with moisture. The phosphine gas is colorless and may ignite spontaneously on contact with air. It has a garlic odor and may have a fishy smell. It inhibits cytochrome oxidase and cellular metabolism and can result in pneumonitis and pulmonary edema. It can also decrease the systemic vascular resistance (mechanism unknown) and cause hypotension.

Polymer Fume Fever

Pyrolysis of fluorocarbons, particularly tetrafluoroethylene, has been reported to cause a condition with symptoms very similar to metal fume fever.

METHEMOGLOBIN PRODUCERS

There are many miscellaneous methemoglobin producers. Copper sulfate is an emetic and a fungicide. Monolurion is a urea herbicide. Naphthalene is found in mothballs and deodorizers. The oxides of nitrogen are found in silage and have been reported to cause methemoglobin from smoke inhalation, particularly from burning plastics. Paraquat (an herbicide), permanganate salts (folk remedy), phenol derivatives (disinfectants), and toluidine (dyes used in the manufacture of artificial fingernails) have been reported to cause methemoglobinemia.

Nitrates/Nitrites

Nitrites are often added to food as a preservative because they can prevent the formulation of *Clostridiumbotulinum* spores. Nitrites are also added as food coloring agents to fix the red color in meat and have been approved for this purpose by the USDA.

The concentration of nitrates is often very high in foods such as cauliflower, spinach, and broccoli, but they rarely cause methemoglobinemia because they are not strong oxidizers. In infants, nitrates are converted in the gastrointestinal tract and are more likely to cause methemoglobinemia because of the *E. coli* found in the upper gastrointestinal tract of infants and the lower gastric acidity. High methemoglobin levels in infants have been reported back to the mid 1960s. Additionally, there have been two fatal cases of nitrite poisoning from food that had been contaminated during transportation by leaking cooling fluid that contained sodium nitrite as an anticorrosive agent.

Nitrates and nitrites are readily absorbed through the skin, mucous membranes, respiratory system, and gastrointestinal tract. Once absorbed and in the bloodstream, these nitrogen compounds combine with hemoglobin and change the ferrous (Fe++) iron component of hemoglobin into ferric (Fe+++) iron-forming methemoglobin. Since the ferric form is unable to carry oxygen, hypoxia results. This condition of hypoxia is called *methemoglobinemia*. The color of the blood while in this condition is chocolate brown and is easily seen during the drawing of blood.

Symptoms of exposure reflect the vasodilating actions of nitrates and nitrites. Fullness in the head and throbbing headache are caused by dilation of meningeal vessels. Flushing of face and neck are signs of cutaneous vasodilation. Dizziness and syncope (swooning or fainting) are due to cerebral ischemia caused by hypotension. Tachycardia, sweating, and pallor are manifested as a response of the sympathetic nervous system to hypotension. Methemoglobinemia causes symptoms such as cyanosis that will not lessen with oxygen therapy. Other symptoms include headache, dizziness, and tachypnea.

Chlorates

Chlorates are used in the manufacture of explosives, dyes, and herbicides. The chlorates are a white, crystalline powder sometimes mistaken for sugar. Of the two forms of chlorates—sodium and potassium salts—the potassium form is more lethal. Chlorates cause gastrointestinal symptoms: nausea, vomiting, diarrhea, and abdominal pain. They can cause methemoglobinemia, intravascular hemolysis, and acute renal failure. The sodium chlorate is freely dialyzable. It is amenable to hemodialysis or exchange transfusion. Sodium thiosulfate, either orally or intravenously, actually inactivates the chlorate ion and is, potentially, a useful antidote. The chlorate methemoglobinemia seems to be particularly resistant to methylene blue.

Nitrobenzene

Nitrobenzene is used in the production of aniline and in the manufacture of polyurethanes. It is also used in the synthesis of acetaminophen. At room temperature it is a liquid and has the smell of bitter almonds, similar to the smell of cyanide. Nitrobenzene is reduced to aniline in the intestine. It is absorbed well by the skin. There are no case reports of toxicity from the lungs or the skin. After ingestion, symptoms may be evident after 30 minutes to 12 hours.

Aniline

As far back as 1886, dyed shoes and dye-stamped diapers containing aniline dyes were being reported to cause methemoglobinemia. Today aniline is used in dyes, pharmaceuticals, photography, polish, resin, varnish, and perfumes. It is a colorless liquid that is highly absorbed by inhalation and through the skin. Aniline itself does not produce methemoglobinemia, but it is metabolized to active compounds that are further oxidized to nitrosobenzene. In addition to the methemoglobinemia, the aniline dyes can cause a hemolytic anemia. Pyridium, which is a therapeutic agent, is metabolized to aniline and aniline metabolites. One case report has shown a methemoglobinemia secondary to the topical application of an aniline cocaine product.

CORROSIVES

The third major category of health hazards is corrosives. Both acids and bases will cause severe damage to living tissue when brought in contact with it. A *corrosive* is a liquid or solid that causes visible destruction or irreversible alteration in human skin tissue at the site of contact. In the case of leaking from its packaging, a corrosive may also be a liquid that is severely corrosive on steel. Some substances meet the definition of corrosives but are classified as other hazards. For example, acid hydrogen peroxide is a corrosive but is classified and shipped as an oxidizer.

Numerous hazards are presented by corrosives. At times, the vapor given off from a leak is so penetrating that workers must wear special acid suits and breathing apparatus to protect themselves.

A second problem to be encountered is that acids such as nitric acid act as an oxidizing agent, and when brought into contact with products containing cellulose, there is a possibility of spontaneous ignition. Large amounts of heat are produced when water is added to sulfuric acid, so one must be assured of the exact compound before deciding to dilute an acid spill with water. In addition to being corrosive, some acids are also extremely poisonous. Hydrocyanic acid, a fairly weak acid, is a very strong poison.

The organic acids are flammable, so in addition to concern for health aspects of the incident, emergency response personnel must ensure that sources of ignition are removed. Inorganic acids can also oxidize and cause ignition in nearby combustibles. Some can act as explosives. Peracetic acid, for example, is extremely dangerous when exposed to heat, and becomes heat and shock sensitive.

Hydrofluoric Acid

Hydrofluoric acid (HF) is one of the strongest inorganic acids known. Because of its prevalence in industry and its wide use, the probability of injuries is great. Hydrofluoric acid is also unique in the way it produces injury. First, it causes a corrosive burn to the skin. The fluoride ion penetrates the skin and causes further injury to the underlying structure.

Acute exposure to hydrogen fluoride will result in irritation, burns, ulcerous lesions, and necrosis of the eyes, skin, and mucous membranes. Total destruction of the eyes is possible. Other effects include nausea, vomiting, diarrhea, pneumonitis, and circulatory collapse. Acute exposure to hydrogen fluoride may require decontamination and life support for the victims. Emergency personnel should wear protective clothing appropriate to the type and degree of contamination.

Hydrofluoric acid tends to remain associated (1,000 times more so than hydrochloric acid) and, as a result, penetrates further into the tissues, often as far as the bone. Depending on the amount of material, symptoms may be delayed, often as long as 24 hours. When sufficient product is present, the fluoride ion separates (dissociates) from the hydrogen. The fluoride ion will then combine with other ions such as calcium or magnesium. This removes those ions from the system as far as being able to accomplish their function. Results of this invasion can be bone demineralization and necrosis. Potassium is released from the nervous tissue as a result of the calcium becoming ineffective and intense pain occurs. If the exposure is to large amounts of product, hypocalcemia may occur, followed by cardiac arrhythmias.

SOLVENTS

Some basic principles that apply to most of the solvents need to be discussed. Most solvents are measured in parts per million, which is a function of room temperature and atmospheric pressure. The American Conference of Governmental Industrial Hygienists and OSHA have developed threshold limit values for exposure including the time-weighted average, the short-term exposure limit, and the ceiling at which patients can be exposed. These numbers are helpful in sorting out potential toxicity. Most solvents are well absorbed by the lungs (perhaps as high as 80%) and have variable solubility through the skin; therefore, exposure values are very indicative of absorbed dose. Since the primary toxicity is absorption through the lungs, the chemicals will circumvent the liver extraction process and go directly to target organs, including the central nervous system and the heart, where they cause their greatest toxicity acutely. Peak blood levels after inhalation absorption occur quickly, within 15 to 20 minutes. Some of the solvents are metabolized to toxic products, and some of the metabolites excreted are measurable in the urine. Most are eliminated (excreted) to a large degree by the lungs, which makes them amenable to breath analysis. Breath analysis is used clinically today in alcohol as well as carbon monoxide exposures. Some research has been done on the use of infrared and breath decay curves.

Acute symptomatology of solvent exposure is fairly universal. Asphyxiation can occur. By reducing the O_2, the patient runs the risk of having relative hypoxia. The major universal toxicity is CNS. There is usually initial euphoria and hyperactivity that turns into depression, coma, and seizures. The solvents are able to cause heart arrhythmias. Through an unclearly defined mechanism, there is increased sensitization of the cardiac cell to circulating catacholamines. It is not uncommon for patients to die from an exposure that has been followed by a fright syndrome. There may be a cardiovascular depression with decreased contractility and cardiac output. (This is considered to be, potentially, reversible by calcium administration. Most of the solvents have pulmonary irritating effects that are, generally, minimal, but some can cause significant pulmonary edema. If ingested, the gastrointestinal symptoms are nonspecific; however, there is a greater tendency to develop hepatitis. The primary compound, or its metabolites, may cause acute tubular necrosis or peripheral neuropathy. An interesting phenomenon called "degreaser's flush" has been reported. The mechanism is unclear, but when patients who have been exposed to solvents ingest alcohol, they may develop a diffuse flush of the skin.

Benzene

Benzene is used mostly as an intermediate in the manufacture of chemicals. Its use as a solvent has decreased significantly because of its toxicity. Approximately 2% of the total amount used is still used as a solvent. Benzene is found in gasoline in the United States at less than 1%. In some other countries, it may be as high as 5%. Obviously, it occurs in automobile exhaust. It has a significant toxic metabolite, benzene oxide, which accumulates in the bone marrow and fat.

Symptoms of lightheadedness, headache, nausea, euphoria, and unconsciousness can be brought on by exposures to concentrations as low as 3,000 ppm. Individual responses will be dramatically different. Concentrations approaching 20,000 ppm will rapidly cause death. As with any product, the length of exposure and the concentration are significant in the results. Exposures to concentrations as low as 7,500 ppm for 30 minutes is considered life-threatening. Long-term exposure affects the blood and immune system. After a latency period of several years, these patients may develop cancer such as stem cell acute myologic leukemia, aplastic anemia, or a multiple myeloma. This toxicity is usually preceded by a leukopenia and thrombocytopenia. If ingested, activated charcoal might be of help since it is known to absorb benzene.

Toluene

Toluene is a component of gasoline added to improve the octane rating and is also used as a solvent. Toluene is found in the gases from a variety of products found in the home such as solvents, paints, and plastics. It is also a component of tobacco smoke. It generally has replaced benzene in many products because it is less volatile and less carcinogenic. Toluene degrades very rapidly in air and, when mixed with water, volatilizes to air very rapidly. It is not as well absorbed as benzene is, either through inhalation or skin. It definitely is not as well absorbed through the skin as phenol. Toluene is rarely absorbed in sufficient quantities to produce a problem. The most common route of exposure is inhalation. Environmentally, gasoline fumes are the most common source. Indoor pollution plays a role, and tobacco smoke provides a significant amount. Acute inhalation exposure, particularly in glue sniffers, may cause loss of consciousness. Distal renal tubular acidosis with a metabolic acidosis, anion gap, and hypokalemia can occur and are usually reversible within 72 hours after stopping exposure. Chronic exposures can cause permanent cerebellar ataxia.

Xylene

Xylene is used as a degreaser and in aviation fuel. Hospitals and research laboratories such as those found in colleges and universities will have significant amounts because it is used extensively in histology laboratories. It is also used as a starting material for the manufacture of polymers and has been used as a solvent for pesticides. It is generally more toxic in acute exposure than benzene and toluene. Xylene is irritating to the mucous membranes and the respiratory tract and causes reversible renal failure and mild hepatitis. Ethanol, when taken concurrently, will inhibit the metabolism of xylene. Xylene has anesthetic properties at concentrations above 5,000 ppm. Short-term exposures to concentrations as low as 300 ppm have been shown to reduce short-term memory and to increase reaction times for persons performing physical activities.

Phenol

Phenol is found in many household items in low concentrations, from disinfectants to chemicals used for home facials. It is used in facials as a "controlled burn" agent where the top layer of skin (corneum stratum) is actually burned away, leaving new pink younger-looking skin. Phenol is carbolic acid and was first introduced as a disinfectant. Chronic phenol toxicity first appeared in medical personnel in the late 1800s when 5% to 10% phenol was used as a skin disinfectant. Phenol is used today in disinfectants, insecticides, and in outpatient surgery by podiatrists. Even an over-the-counter medication, Campho-Phenique, contains 2% to 5% phenol.

Phenol denatures proteins found in the skin through a process of *saponification,* which converts fat into soap and leaves a whitish or brownish discoloration on the skin. It has the ability to be absorbed by any surface of the body. Once in contact with the skin, phenol causes intense pain and blanching of the area. Phenol is absorbed rapidly through the lungs and through the skin. Symptoms develop in 5 to 30 minutes. Even low concentrations can cause devastating results. There are reports of children dying after application of compresses with 5% phenol solutions (see Table 10.4).

A recent case report describes a situation in which an outpatient was mistakenly given one ounce of 89% phenol solution by mouth. She immediately clutched her throat and lost consciousness. Within 30 minutes she had no recordable blood pressure and sustained respiratory arrest. During intubation, the patient's mouth and pharynx were reported to be white. A "lamp oil" odor was noted on her breath while being ventilated. The patient experienced ventricular tachycardia and was converted with cardioversion. She continued with ventricular ectopy (a change in heartbeat rhythm), seizures, and metabolic acidosis, but eventually recovered after 15 days.

The systemic effects of phenol toxicity are evidenced through central nervous system depression with coma and respiratory arrest. An initial excitation phase may be seen early on. Convulsions are common and have appeared as late as 18 hours after exposure.

Treatment of phenol poisoning includes decontamination of the skin using copious amounts of water, followed by olive oil or isopropyl alcohol for at least 10 minutes. If only small amounts of water are used to decontaminate, phenol absorption

▶ **TABLE 10–4**
Phenol-related compounds.

EXAMPLES OF PHENOL RELATED COMPOUNDS	
Amyl phenol	Germicide
Creosol	Antiseptic
Creosote	Wood preservative
Quaiacol	Antiseptic
Hexachlorophene	Antiseptic
Medicinal tat	Treatment of dermatological conditions
Phenol	Outpatient podiatric surgery
Phenylphenol	Disinfectant
Tetrachlorophenol	Fungicide
Thymol	Fungicide and anthelmintic

will increase and make the patient's condition worse. Supportive measures such as assisting ventilations and controlling ventricular ectopy and seizures are done as the need arises. Continuous cardiac vital sign monitoring and application of 100% oxygen are mandatory.

Hexane

The straight-chain hydrocarbons are referred to as *aliphatics*. If there are less than four carbons (such as methane, ethane, propane, and butane), these hydrocarbons exist as gases. Hydrocarbons that have 5 to 16 carbons are liquids. Those with more than 16 carbons are waxes. Hexane is a six-carbon aliphatic and is used as a thinner in rubber and in the food, drug, and perfume industries. It has a major metabolite that is toxic and inhibits the energy for nerve protein transport leading to degeneration of the nerve. After a two- to six-month delay, the patient may develop a reversible sensory neuropathy that progresses to a motor neuropathy. Chronic exposure can lead to changes in color vision and permanent CNS changes.

Gasoline is made up of a mixture of aliphatics. It boils between 40°C and 250°C (104°F and 482°F). It also contains xylene, tetraethyl lead, and some alcohols. Kerosene is used as a charcoal lighter fluid, solvent, or fuel. It is a mixture with a boiling point between 175°C and 325°C (347°F and 617°F) and contains a mixture of straight chains, xylene, and naphthalene. Diesel oil and diesel fuel are made up of mixed chains. Mineral spirits, which are used as a dry cleaner and a solvent and paint thinner, contains straight chains, naphthalene, and some aromatics. Naphtha, which is a lighter fluid, thinner, and a varnish, consists of mixed chains and boils between 94°C and 175°C (201°F and 347°F). Turpentine is not truly a petroleum distillate, but is a steam distillation of pine resin.

Halogenated Aliphatics

Chloroform is a less potent hepatotoxin than carbon tetrachloride. Tetrachloroethylene is a common dry cleaning and textile industry chemical that accounts for approximately two-thirds of its use. Trichloroethylene is an automotive and metal degreaser and, prior to 1977, was used as an anesthetic agent and an analgesic in medical practice. It is still used in veterinary medicine. It has also been used to decaffeinate coffee; however, methylene chloride is more commonly used for that purpose today. Ethylene dichloride is used in the manufacture of PVC plastic and as a lead scavenger in gasoline. It causes some olfactory paralysis similar to hydrogen disulfide. It is a hepatorenal toxin and can cause a particular bluish-purple discoloration of the skin, dermatitis, and corneal abrasions. Trichloroethane is used as a degreaser, in hairsprays, cosmetics, oven cleaners, spot removers, and as a propellant in some decongestant sprays. Trichloroethane is the cause of 28% of sudden deaths in glue sniffers by one study and has been reported to cause Goodpasture's syndrome, acute glomerulonephritis, and pulmonary infiltrates. Propylene dichloride is used as a solvent and stain remover and has been reported to cause renal failure and hemolytic anemia, as well as bluish-purple discoloration of the skin.

Carbon Tetrachloride

Carbon tetrachloride was used historically in fire extinguishers and as a medicinal to get rid of worms. Because of its recognized toxicity, it is primarily used today only in industry and as a solvent. Chloroform is a natural metabolite. Fire produces phosgene and hydrochloric acid, which is why in the 1960s carbon tetrachloride was banned for use in fire extinguishers. Carbon tetrachloride causes an acute fatty degeneration of the liver, acute tubular necrosis of primarily the proximal tubule and the loop of Henle in the kidneys, pulmonary hemorrhage and edema, and CNS depression that resembles encephalitis with particular cerebellar vulnerability. Hyperbaric oxygen has been used to treat carbon tetrachloride exposure; however, there is concern that it may increase the development of phosgene. N-acetylcysteine has been used in the same dose as that of acetaminophen to reduce the reactive metabolites and prevent hepatorenal toxicity.

Methylene Chloride

Methylene chloride is primarily used as a paint remover and flammability depressant. It is a degreaser, and it is also used to extract caffeine from coffee. The principal exposure is through inhalation; it causes minimal CNS effects from oral or dermal exposure. Methyl chloride is metabolized to produce carbon monoxide, although it is generally considered that the carboxy-hemoglobin level will maximally reach 20% and is not generally high enough to cause the CNS effects. The combination of the methylene chloride and the carboxyhemoglobin may do so, however. It is excreted in the breath, and the liver and CNS are the primary targets of toxicity. As is true with almost all solvents, patients can develop headache, dizziness, nausea, paresthesias, and loss of consciousness. Management is primarily with oxygen and EKG monitoring. Although there are not studies to suggest such, the use of hyperbaric oxygen has also been suggested in acute, severely toxic patients.

Alcohols

Alcohols are a broad class of hydroxyl-containing organic compounds. They occur naturally in plants and are produced synthetically from petroleum derivatives. Alcohols are used in organic synthesis, as solvents and detergents, in beverages, in pharmaceuticals and plasticizers, and as fuels. Of particular interest are methanol and ethylene glycol. Methanol is found in windshield washer fluid, is metabolized to formaldehyde, and is quickly metabolized to formic acid. Methanol and, to a certain degree, formic acid cause CNS depression. The acidosis and retinopathy are probably secondary to the formic acid and formaldehyde. Ethylene glycol, which is commonly found in radiator fluid, is also metabolized to an aldehyde, then to a glycolic acid and an oxalic acid. The primary mechanisms of toxicity are acidosis and renal failure. Isopropyl alcohol is metabolized to acetone and causes increased serum acetone, urine acetone, and minimal acidosis. Methanol and ethylene glycol exposure is treated by using ethanol to block an enzyme that prevents metabolism and to hemodialyze significant exposures.

Ketones

Ketones, as a group, are solvents that are used in resins, lacquers, and dyes and in the production of cotton and various pigments. Methyl-n-butyl ketone metabolizes to 2,5-hexanedione and causes peripheral neuropathy. Methyl ethyl ketone is a solvent that causes contact dermatitis and irritation of the mucous membranes. Acetone is found in nail polish remover, varnish, and glues and is the by-product of fat metabolism in diabetic ketoacidosis. Acetone is exhaled unchanged; however, the majority is eliminated by the kidney.

Carbon Disulfide

Carbon disulfide is a highly volatile liquid that smells like chloroform. The vapor is extremely hazardous and is usually inhaled; however, it can also be absorbed. It is used as solvent for oils, fats, and rubber and as a corrosion inhibitor, insecticide, and wax. It is an excellent fat solvent. It can also be used as a disinfectant, a fumigant, and in the manufacture of rayon and artificial silk. It has the odor of decaying cabbage and is rapidly absorbed when inhaled. Once absorbed, most of it is metabolized; however, some is stored in body fat. The principal effect is CNS stimulation followed by headache, nausea, and vomiting, fatigue, excitement, euphoria, and hallucination, with death secondary to respiratory failure. Disulfiram (Antiabuse) is metabolized to carbon disulfide, which can deactivate the acetylaldehyde dehydrogenase needed for ethanol (grain alcohol) metabolism. Chronic exposure leads to neuropsychiatric disturbances, neuropathies, and increased heart disease.

TOXICITY OF METALS

Lead

Lead is ubiquitous in our environment, and its toxic preparations have been written about for centuries. A Greek physician first described occupational lead poisoning in the second century B.C. There is no known useful purpose for lead in the body, so any that is there is likely to be toxic. Organic lead results from the combustion of gasoline. Vehicle exhaust accounts for as much as 25% of total body lead. This has been reduced from 37% in the 1970s to as low as 25% in the 1980s as a result of the change in the use of leaded gasoline to unleaded gasoline. Inorganic lead is a solid waste product from ore production, iron and steel production, and ammunition use. This is of considerable concern today because of lead shot that is used in hunting, which increases the lead ingested by our wild fowl. Lead solder and lead ceramic glaze may cause water and food contamination and are, therefore, another potential source of contamination. Plumbers who use lead-based solders and old plumbing fixtures may install sources of leachable lead. In a recent study in California, as much as 6.6% of ceramic ware had leachable lead; 67% of the samples tested were handmade in Mexico. Current legislation limits the lead content of paint to less than 0.06%. Homes built before 1940 may contain concentrations of lead-based paint as high as 50%; houses built before 1978 are considered at risk. Between 40% and 50% of occupied housing in the United States still has lead-based paint on the walls. Children with a pica (an abnormal appetite that is sometimes associated with low levels of iron, calcium, and zinc) may have an increased lead absorption that may be higher than 40% of what is considered normal. Some folk remedies, such as those used in Hispanic, Hmong, and Indian folk medicines, as well as eye makeup and gasoline sniffing, may be sources of lead.

If lead is taken orally, the acute symptomatology includes abdominal pain and vomiting and CNS effects leading to encephalopathy in children. Acute encephalopathy is infrequent in adults. Thirty percent of children who have encephalopathy will also have permanent neurological deficits. Subacute symptoms include anorexia and constipation, apathy, memory loss, and arthralgias. Muscle and joint pains are common in lead workers. A peripheral neuropathy is not uncommon, with motor conduction velocities being significantly prolonged; sensory conduction velocities are usually normal. On examination, hypertension, ataxia, wrist drop or foot drop, optic neuritis, papilledema, or purple lines on the gums (lead lines) may be observed.

The form of lead affects its absorption when ingested. Iron and calcium absorption will also affect it. When absorbed, it binds to the red blood cells and is carried throughout the body. When there has been extremely high respiratory exposure, lead poisoning can occur acutely. If the dose is large enough, encephalopathy can develop and may be accompanied by renal failure and gastrointestinal symptoms. Such cases are relatively rare, however, as most of the time, lead is absorbed over a period of weeks or months.

Mercury

Mercury is a general poison. Its major action appears to be the binding to sulfhydryl groups, thereby inactivating certain enzymes. There are three forms of mercury: elemental, inorganic, and organic.

The elemental mercury found in thermometers is volatile and can vaporize. In that form it can be absorbed from the lungs. A broken thermometer needs to be cleaned up as quickly as possible using special equipment. Inorganic mercury is used in dental amalgams, batteries, herbicides, insecticides, and spermicides. Mercury is added to some latex paints as a bactericide and fungicide and has been reported to cause toxicity immediately after application. Organic mercury is found in foods, including fish and livestock that have been fed fungicide-treated grains.

There have been numerous organic mercury poisoning epidemics worldwide. Two major epidemics of methyl mercury poisoning occurred in Japan, one in Minamata Bay and one in Miigata Bay. These resulted from the industrial releases of mercury, conversion by microorganisms, and accumulation in fish. The largest epidemic recorded was in the early 1970s in Iraq in which over 6,000 patients were admitted to hospitals and 500 deaths occurred. This was a result of methyl mercury from bread that was made from seed grain wheat. Seed grain is usually treated with fungicide, in this case, methyl mercury. Similar epidemics occurred in other Third World countries in the 1960s.

Inorganic mercury tends to accumulate in the kidneys and the liver; organic mercury tends to accumulate in the kidneys, the CNS, and the liver and does undergo enterohepatic recirculation. Symptomatology will vary depending on the form of the mercury. Elemental mercury is more likely to cause pulmonary effects, bronchitis, pneumonitis, and some CNS effects such as tremor and hallucinations. The inorganic mercury is more likely to be caustic, particularly if ingested, and is more likely to cause renal dysfunction such as glomerulonephritis. The organic mercuries have early symptoms of sensory neuropathy with paresthesias and ataxia, progressing to a central neuropathy or a coma. The classic triad of symptoms is dysarthria, ataxia, and visual field constriction. Mercury poisoning may, in fact, mimic ALS (amyotrophic lateral sclerosis). The whole blood is probably the best measure of recent exposure to inorganic or elemental mercury. Urine levels have been used to assess exposure to inorganic mercury. Activated charcoal seems to interrupt the absorption of inorganic mercury. Acute toxicity is usually treated with BAL, an antidote of dimercaprol. Dimethyl sulfoxide (DMSO), which is an oral chelator, and oral resins that are nonabsorbable may chelate mercury within the gut. Hemodialysis seems to be of limited value.

Arsenic

Arsenic, in the metallic form, is used as an alloying additive for metals, especially lead and copper as shot, battery grids, cable sheaths, and boiler tubes. It is used to make gallium arsenide for dipoles and other electronic devices. It is found in coal-burning power plants, smelters, insecticides, herbicides, fungicides, and algicides. Agricultural products account for approximately 80% of arsenic use. It is highly corrosive and can cause hemorrhagic gastritis leading to shock through dilation of vessels. Development of cerebral edema leads to encephalopathy with coma and convulsions. It can cause jaundice and renal failure. The few tip-offs of acute exposure may be intense abdominal pain, bloody diarrhea from the caustic effects, or a garlic odor (which, as has been pointed out, can represent a number of toxins). Once the diagnosis of acute intoxication has been noted, a chelator (BAL or dimercaprol) may be used. Chronic exposure may lead to a sensory neuropathy, a condition of gangrene called Blackfoot's disease, and squamous cell carcinoma called Bowen's disease. As much as 7% of all skin cancers may be due to arsenic exposure.

Chromium

Hexavalent chromium compounds are carcinogenic and corrosive on tissues, which may result in ulcers and dermatitis with prolonged contact. Chromates are used in a number of industries. Absorption is high if ingested through the gastrointestinal tract or through the lungs via vapor. Chromates may be absorbed through the skin, even if it is intact. Chromic acid may cause severe burns or erosion of the skin. Allergic contact dermatitis may occur. Accidental ingestion has occurred in a worker exposed to the chemical used as a radiator de-scaler. Chelation is of little value. Dialysis and exchange transfusion should be considered as treatment for exposure.

Manganese

Manganese is found in steel, batteries, ceramics, glass, and dyes. The germinal portion of certain grains, fruits, nuts, and tea are very high in manganese. Acute exposure to manganese may produce metal fume fever, a flu-like syndrome. It readily crosses the blood-brain barrier and has been known to cause a condition referred to as manganese madness (confusion, hallucinations, difficulty with speech and fine motor control, and loss of balance). Chelation therapy with EDTA might be of value in managing this acute psychosis.

Thallium

Thallium is used in certain pigments and thermometers and was widely used as a medicinal agent for venereal disease, ring worm, gout, dysentery, and tuberculosis in the 1800s. During the middle part of this century, it was used in a rodenticide, but was banned in 1965. It causes gastrointestinal effects, central nervous system effects, and a motor sensory neuropathy. Thallium is excreted in the bile. For that reason, poisoning may be amenable to repeat doses of activated charcoal. Chelation therapy, particularly with dithiocarb, may cause a redistribution in the CNS and worsen the symptoms.

Copper

Copper is found in beverages and in some drinking vessels, particularly those that are brought in from outside the United States. This can have primarily gastrointestinal effects that can produce a vomitus that is greenish-blue. Patients can develop a hemolytic anemia, renal failure, and coma.

Selenium

Selenium is used in electronics, glass, ceramics, and steel and is found in a product called gun bluing, which is a product used to polish the metal of a gun. It is highly absorbed through the lungs and the gastrointestinal tract. Like phosphorus compounds and arsenic, it has a garlic-like odor. It can cause a toxic cardiomyopathy resulting in decreased contractility and shock, respiratory depression, stupor, and coma. Chelation may enhance toxicity. There are no antidotes. The chronic picture resembles that of arsenic poisoning.

Tin

Organic tin has been used medicinally in France. The organotins are used as additives in many products and processes that require preservation. Tributyltin oxide is used for mildew and fungicide control by including it in paints. It is a potent CNS system toxin causing headaches, psychosis, and seizures. The organotins are also known to have immunotoxic properties in animals.

SUMMARY

▶ Blast injuries produce serious injuries and trauma. The effects from the blast wave cause injury not only to the surface of the body but also to the internal organs.
▶ Irritant gases cause damage primarily to the lungs. Some of the major irritant gases that can cause injury are acrolein, sulfur dioxide, hydrogen chloride, ammonium nitrate, ammonia, ozone, and fluorocarbons.
▶ Toxic gases can cause both lung and systemic injuries. Some toxic gases include carbon monoxide, cyanide, and hydrogen sulfide.
▶ Methemoglobin producers include nitrates and nitrites and chlorates.
▶ Corrosives, such as hydrofluoric acid, can cause severe damage to any living tissue brought in contact with it.
▶ Solvents and metals, such as chromium, can also cause many harmful effects in our environment.

STUDY QUESTIONS

1. What are the main effects of blast wave injuries?
2. Identify three irritant gases and describe the toxic effects of each.
3. What are the differences between irritant gases and toxic gases?
4. Identify three irritant gases and describe the toxic effects of each.
5. What are the symptoms of methemoglobinemia?
6. Identify three solvents and describe the toxic effects of each.
7. What are some of the ways people can be exposed to toxic levels of metals? How can such exposures be mitigated?

Appendix A
Chemicals in the Workplace

Chemical	Effects
Acetaldehyde (CH_3CHO)	General narcotic action. Large dose may cause death by respiration paralysis.
Acetaldol ($C_4H_8O_2$)	Irritates skin and eye tissues
Acetic acid glacial (CH_3COOH)	Ingestion may cause severe corrosion of mouth and G.I. tract.
Acetone (CH_3COCH_3)	Prolonged or repeated topical use may cause erythema, dryness. Inhalation may produce headache, fatigue, and bronchial irritation.
Acetophenone (C_8H_3O)	Irritates eyes and respiratory tract, also mutagenic
2-Acetoxyacrylonitrile ($C_4H_5O_2CN$)	Toxic by ingestion, inhalation, and skin absorption
Acetyl bromide (CH_3COBr)	Very irritating to eye tissues
Acetyl chloride (CH_3COCl)	Very irritating to eye tissues
Acetylene (H_2C_2)	A simple asphyxiant. High concentrations cause narcosis.
Acrolein (C_2H_3CHO)	Highly toxic. Can severely damage eyes and respiratory system
Acrylic acid ($CH_2=CHCO_2H$)	Corrosive liquid, strong irritant
Acrylonitrile (C_2H_3CN)	Highly toxic. Carcinogenic and teratogenic
Alachlor ($C_{14}H_{20}ClNO_2$)	Mutagenic
Aldicarb ($C_7H_{14}N_2O_2S$)	Highly toxic pesticide
Aldrin ($C_{12}H_8Cl_6$)	Headache, nausea, vomiting, convulsions. Chronic doses lead to liver disease.

Allyl alcohol (C_3H_6O)	Causes severe irritation of mucous membranes and eyes
Allyl chloride (C_3H_5Cl)	Produces irritation of eyes and respiratory passages. Readily absorbed through skin
Ammonia (NH_3)	Inhalation of concentrated vapor causes swelling of respiratory tract tissues, spasm of the glottis, and asphyxia.
Ammonium bisulfide (NH_4HS_2)	Very irritating to skin. Penetrates very rapidly and may be fatal
Ammonium fluoride (NH_4F)	Ingestion produces nausea, vomiting, diarrhea, convulsions, vascular collapse, and death.
Ammonium hydroxide (NH_4OH)	Highly corrosive. Very irritating to skin and eye tissue
Aniline ($C_6H_5NH_2$)	Moderately toxic. May be carcinogenic
Antimony (Sb)	Irritant of respiratory membranes. Irritates skin
Arsenic (As)	Most forms are very toxic to humans.
Arsenic acid (H_3AsO_4)	Toxic because the acid converts to As_2O_3 by heating above 300°C
Arsine (AsH_3)	Poisonous gas, carcinogenic
Asbestos	This calcium magnesium silicate with occupational exposure to dust can result in cancer (mesothelioma).
Atropine ($C_{17}H_{23}NO_3$)	An antidote for several toxic substances. Overdose can cause severe poisoning.
Barium azide ($Ba(N_3)_2$)	Explosive
Barium carbonate ($BaCO_3$)	Toxic (rat poison). Causes vomiting, violent diarrhea, convulsive tremors, and muscular paralysis
Barium chlorate ($BaCl_2O_6$)	Very poisonous
Barium hydroxide ($Ba(OH)_2$)	Highly corrosive, very irritating to skin and eye tissue
Benzene (C_6H_6)	Irritation of mucous membrane, restlessness, convulsions. Death may follow from respiratory failure. Causes chronic depression and leukemia
Benzidine ($C_{12}H_{12}N_2$)	This chemical, which is used in manufacturing dyes, has been declared a carcinogen by the FDA.
Benzoic acid ($C_7H_6O_2$)	Mild irritant to skin and eyes
Benzyl chloride (C_7H_7Cl)	Corrosive liquid that irritates the skin, eye tissue, and throat
Beryllium (Be)	Death may result from short exposure to very low concentrations of powdered beryllium and beryllium salts.
Bis (chloromethyl) ether ($(CH_2Cl)_2O$) also called BCME	Highly toxic and carcinogenic
Boric acid (H_3BO_2)	Ingestion or absorption may cause nausea, vomiting, diarrhea, or circulatory collapse.
Boron trifluoride (BF_3)	Corrosive to skin, avoid inhalation
Bromic acid ($HBrO_3$)	Highly irritating to skin, eyes, and mucous membrane
Bromine (Br)	Burns and blisters the skin, very irritating to respiratory tract. Bromine compounds can be very caustic.
Bromine azide (BrN_3)	Explosive
Bromine pentafluoride (BrF_5)	Very corrosive to skin, eye tissue, and mucous membranes

Bromobenzene (C_6H_5Br)	Irritating to skin
Butanol (C_4H_9OH)	High concentrations are teratogenic (cause birth defects) in rats.
Butyl acetate ($C_6H_{12}O_2$)	Irritates eye tissue, nose and throat. Chemical produces a narcotic effect.
2-Butyne-1,4-Diol ($C_4H_4(OH)_2$)	Moderate to high toxicity in test animals. Irritant to skin
Butylamine ($C_4H_9NH_2$)	Irritant to skin, eye tissue, and respiratory tract
Butyric acid ($C_4H_8O_2$)	Corrosive, flammable liquid. Ingestion causes vomiting and diarrhea. Convulsions occur at higher concentrations.
Cadmium (Cd)	Powdered metal is toxic, and soluble compounds are toxic.
Cadmium cyanide (C_2CdN_2)	This electroplating chemical is very poisonous.
Calcium carbide (CaC_2)	With water produces flammable acetylene
Calcium hydroxide ($Ca(OH)_2$)	Moderate irritation to skin and eye tissue
Carbon disulfide (CS_2)	Nausea, vomiting, convulsions. Long-term exposure causes psychic disturbances.
Carbon monoxide (CO)	Combines with hemoglobin and causes dizziness, nausea, vomiting, decreased pulse and respiratory rates, and death
Carbon tetrachloride (CCl_4)	Poisonous by inhalation. Ingestion or absorption by skin can be fatal.
Chloral cyanohydrin ($C_2(OH)Cl_3CN$)	Toxic response in laboratory animals observed
Chloral hydrate ($Cl_3C_2H_3O$)	Causes general anesthesia. High doses may be lethal.
Chlordane ($C_{10}H_6Cl_8$)	Highly toxic, nausea, vomiting, convulsions, high doses fatal
Chloric acid ($HClO_3$)	Highly irritating to skin, eyes, and mucous membranes
Chlorine (Cl)	Can cause fatal pulmonary edema. Chlorine compounds may be very caustic.
Chlorine azide (ClN_3)	Explodes spontaneously
Chlorine pentafluoride (ClF_5)	Highly toxic gas, irritant to skin, eye tissue, and respiratory tract
Chlorine trifluoride (ClF_3)	Irritant to skin, eye tissue, and mucous membranes
Chloroacetaldehyde ($ClCH_2CHO$)	Highly toxic and corrosive to eyes, skin, and respiratory system
Chloroform ($CHCl_3$)	Carcinogen
Chromium (Cr)	Chromic acid and chromic salts are irritants to human skin and membranes.
Cinnamaldehyde (C_9H_7ClO) or (C_9H_6BrClO)	A group of chemicals with a basic aldehyde structure that has an additional chlorine or bromine group. Mutagenic and also causes muscle tremors and contractions as well as irritant to skin
Cocaine ($C_{17}H_{21}NO_4$)	Low doses euphoria. High doses convulsions, hypothermia, and respiratory failure
Codeine ($C_{18}H_{21}NO_3$)	Habit-forming. An overdose produces respiratory failure.
Coniine ($C_{18}H_{17}N$)	Highly toxic plant poison from hemlock which causes convulsions, asphyxia, and death
Creosote ($C_8H_{10}O_2$) (mixture)	Large doses may cause G.I. irritation, cardiovascular collapse, death.

Crotonaldehyde (C_3H_5CHO)   Causes severe irritation of eyes and respiratory system

Cumene (C_9H_{12})   Irritant of eyes, skin, and mucous membranes

Cupric arsenite ($CuHAsO_3$)   Very poisonous wood preservative.

Cuprous fulminate ($CuCNO$)   Explosive

Cyanic acid ($HOCN$)   Highly irritating to skin, eyes, and mucous membranes

Cyanide (CN)   Cyanide compounds are to be considered highly toxic in most forms.

Cyclohexanol ($C_6H_{11}OH$)   Toxic to eyes, skin, and respiratory tissues

Cyclohexylamine ($C_2H_{11}NH_2$)   Irritant of skin, eyes, and respiratory tract

DDT ($C_{14}H_9Cl_5$)   Poisoning may occur by ingestion or by absorption through skin or respiratory tract.

2,4-D ($C_8H_6C_{12}O_3$)   Nausea and vomiting. At high doses causes convulsions and coma

1,4-Dichlorobenzene (C_6H_4Cl)   Irritant of skin, eye tissue, and throat. May be carcinogenic

Diethylamine ($C_4H_{10}NH$)   Irritant to skin, eye tissue, and respiratory tract

Diethyl ketone ($C_5H_{10}O$)   Irritant and mild narcotic

Digitalis ($C_{36}H_{56}O_{14}$)   Cardiotonic

Dimethylamine (C_2H_6NH)   Irritant of skin, eye tissue, and mucous membranes

2,4-Dinitrophenol ($C_6H_4N_2O_5$)   Nausea, vomiting, and death

Dioxane ($C_4H_8O_2$)   CNS depression, necrosis of liver and kidneys

Dioxin ($C_{12}H_6Cl_2O_2$)   A class of related chemicals with many different chlorine groups attached; very toxic, carcinogenic, and teratogenic

Diquat dibromide ($C_{12}H_{12}N_2Br_2$)   Herbicide that causes respiratory distress

EPN ($C_{14}H_{14}NO_4PS$)   Cholinesterase inhibitor, insecticide (Parathion)

Ethanol (C_2H_5OH)   Ingestion of large doses (250 ml) in a short time can be fatal

Ethyl acetate (C_4H_8O)   Irritant to eye tissue, nose, and throat

Ethylamine ($C_2H_5NH_2$)   Strong irritant to skin, eye tissue, and respiratory tract

Ethylene chlorohydrin (C_2H_5OCl) Toxic to central nervous system, kidneys, liver, and gastrointestinal tract

Ethylenediamine ($NH_2C_2H_4NH_2$)   Severe skin irritant producing blisters

Ethylene glycol ($C_2H_6O_2$)   Antifreeze. If ingested causes CNS depression, vomiting, respiratory failure, venal damage, death.

Ethylene oxide (C_2H_4O)   Highly irritating to eyes, can cause pulmonary edema

Ethylenimine (C_2H_4NH)   Highly poisonous and a severe irritant to skin, eyes, and mucous membranes

Ethyl ether ($C_4H_{10}O$)   Inhalation causes unconsciousness and death from respiratory paralysis.

Ethyl formate ($C_3H_6O_2$)   Irritant to skin and mucous membranes

Ferric chloride (Cl_3Fe)   Anhydrous form is an irritant to human tissues.

Fluoboric acid (BF_4H)   Caustic to skin and mucous membranes

Fluorine (F)   Very dangerous, caustic to eyes, skin, and mucous membranes. Fluorine compounds may be very caustic.

Fluoroacetic acid ($C_2H_3FO_2$)	Very toxic. Causes convulsions and ventricular fibrillation
Formamide (HCHO)	Moderate to severe injury to eyes, skin, and respiratory system
Formamide (CH_3NO)	Moderately irritating to skin and mucous membranes
Formic acid (HCOOH)	Highly irritating to skin, eyes, and mucous membranes
Furan (C_4H_4O)	Highly toxic vapors can damage lung tissue.
Gasoline (C_4 to C_{12})	Ingestion causes inebriation, vomiting, and cyanosis.
Germane (GeH_4)	Poisonous gas
Glutaraldehyde ($(CH_2)_3(CHO)_2$)	Used to sterilize or disinfect. A strong irritant to eyes and respiratory membranes
Glycidaldehyde ($C_3H_4O_2$)	Irritant to eye tissue, mutagenic
Glycidol ($C_3H_6O_2$)	Irritant to eye tissue, lungs, and skin
Glycolic acid ($C_2H_4O_3$)	Mild irritant to skin and mucous membranes
Glyoxal ($C_2H_2O_2$)	Skin and eye irritant
Heptachlor ($C_{10}H_5Cl_7$)	Highly toxic, liver disease, may be carcinogenic
Heroin ($C_{21}H_{23}NO_3$)	Habit-forming. Overdose produces respiratory failure.
Hexane (C_6H_{14})	Irritating to respiratory tract in high concentrations
Hexanol ($C_5H_{11}CHO$)	Irritant of skin and eye tissue
Hydrazine (H_2N-NH_2)	Highly irritating to eyes, nose, and throat. Highly flammable rocket fuel
Hydrobromic acid (HBr)	Strong irritant
Hydrochloric acid (HCl)	Causes severe burns, coughing, choking, inflammation, ulceration, nausea, death
Hydrofluoric acid (HF)	Severe irritation of eyes. Burns on eyes and skin
Hydrogen azide (NH_3)	Irritation of eyes, cough, headache, fall in blood pressure, collapse of circulation
Hydrogen bromide (HBr)	Irritating to eyes, skin, and respiratory membranes
Hydrogen cyanide (HCN)	Paralysis, unconsciousness, convulsions, respiratory arrest, and death
Hydrogen sulfide (H_2S)	Collapse, coma, death from respiratory failure
Iodine (I)	Elemental form very corrosive
Iodine trichloride (ICl_3)	Very corrosive to skin, eye tissue, and mucous membranes
Iron pentacarbonyl (C_5FeO_5)	Lung irritant, liver and kidney damage
Isobutyl alcohol (C_4H_9OH)	Inhalation causes eye and throat irritation and headache.
Isobutyl alcohol ($C_4H_{10}O$)	Mildly irritating to skin and mucous membranes
Isobutyraldehyde (C_3H_7CHO)	Moderate skin and eye irritant
Isobutyric acid ($C_4H_8O_2$)	Mild irritant
Isopentyl alcohol ($C_5H_{12}O$)	Irritating to mucous membranes
Isoprene (C_5H_8)	Irritating to skin and mucous membranes
Isopropyl acetate ($C_5H_{10}O_2$)	Irritant to eye tissue, nose, and throat
Isopropyl ether ($(C_3H_7)_2O$)	Irritant to skin, eye tissue, and nasal passages. Also has narcotic effect

Isopropyl nitrate ($C_3H_7NO_2$)	Jet fuel. Causes faster heart rate, headache, cyanosis, shock, and death.
Kerosine (C_{10} to C_{16})	Irritation of skin. Inhalation causes headache, drowsiness, and coma. Swallowing results in vomiting and diarrhea.
Ketene (C_2H_2O)	Pulmonary irritant
LSD ($C_{20}H_{25}N_3O$)	Psychedelic agent, nausea and vomiting
Lead (Pb)	Lead dust can cause vomiting, convulsions, and CNS complications. Many lead compounds are also toxic.
Lead azide ($Pb(N_3)_2$)	Explosive, aqueous solution is toxic
Lindane ($C_6H_6Cl_6$)	Poisoning occurs by ingestion or inhalation, with headache, nausea, vomiting, convulsions, cyanosis, and circulatory collapse.
Lithium azide (LiN_3)	Explosive
Lithium bromide (LiBr)	CNS depression; disturbance is blood electrolyte imbalance.
Lithium chloride (LiCl)	Blood electrolyte imbalance, kidney dysfunction
Lithium hydroxide (LiOH)	Caustic to eyes and skin
Manganese (Mn)	Inhalation of dust results in sleepiness, weakness, emotional disturbances, muscle spasticity, and paralysis.
Magnesium perchlorate ($Mg(ClO_4)_2$)	Irritation of skin and mucous membranes
Malathion ($C_{10}H_{19}O_6PS_2$)	Neurological toxin, insecticide
Maleic acid ($C_4H_4O_4$)	Strong irritant
Maleic anhydride ($C_4H_2O_3$)	Powerful irritant, causes burns
Mercuric arsenate ($AsHHgO_4$)	Very toxic
Mercury (Hg)	Toxic-nausea, vomiting, abdominal pain, diarrhea, kidney damage, and death. Many mercury compounds are potentially toxic.
Mercury nitride (Hg_3N_2)	Explosive
Methacrylic acid ($C_4H_6O_2$)	Highly corrosive, can result in blindness
Methanol (CH_3OH)	Causes blindness with ingestion
Methyl acetate ($C_3H_6O_2$)	Irritating to respiratory tract
Methylamine (CH_3NH_2)	Irritating to eyes, skin, and respiratory tract
Methyl bromide (CH_3Br)	Pulmonary edema, CNS depression, kidney damage
Methyl butyl ketone ($C_6H_{12}O$)	Exposure causes disorders of nervous system
Methyl chloride (CH_3Cl)	Inhalation causes headaches, nausea, vomiting at high doses, coma, and respiratory failure.
Methylcyclohexanol ($CH_3C_6H_{10}OH$)	Mildly toxic. Produces irritation to the eyes and respiratory system
Methyl ether $(CH_3)_2O$	Causes sedation, coma, and death
Methyl ethyl ketone (C_4H_8O), also called MEK	Irritant of eyes and respiratory membranes
Methyl formate ($C_2H_4O_2$)	Toxic to eye tissue and respiratory tract
Methyl isoamyl ketone ($C_7H_{14}O$)	Irritant to eyes and respiratory tract
Methyl propyl ketone ($C_5H_{10}O$)	Irritant of eyes and respiratory membranes

Methyl vinyl ketone (C_4H_6O)	Readily absorbed through the skin, causing general poisoning
Morpholine (C_4H_9NO)	Irritating to eyes and mucous membranes
Mustard gas ($C_4H_8Cl_2S$)	Highly toxic and carcinogenic
Naphthalene ($C_{10}H_8$)	Moth repellant results in poisoning by ingestion, inhalation, or skin absorption
Naphthylamine ($C_{10}H_9N$)	Carcinogenic
Naphthylamine ($C_{10}H_7NH_2$)	Moderately toxic and carcinogenic
Nickel (Ni)	Causes dermatitis. Nickel dust causes nausea, vomiting, and diarrhea.
Nickel fluoride (NiF_2)	Chronic exposure causes mottling of teeth and alteration of bone chemistry.
Nickel tetracarbonyl ($Ni(CO)_4$)	Flammable and highly poisonous
Nicotine ($C_{10}H_{14}N_2$)	Concentrated form is highly toxic. The lethal dose is 40mg/kg.
Nitric acid (NHO_3)	Corrosive to eyes, skin, and mucous membranes
Nitrobenzene ($C_6H_5NO_2$)	Absorbed through skin. Causes headaches, nausea, vomiting, and cyanosis.
Nitroethane ($C_2H_5NO_2$)	Irritating to eyes and mucous membranes
Nitroglycerin ($C_3H_5N_3O_9$)	Causes nausea, vomiting, paralysis, convulsions, circulatory collapse, and death
Nitropropane ($C_3H_7NO_2$)	Irritating to mucous membranes
Nitrosophenol ($C_6H_5NO_2$)	Skin irritation, explodes on contact with acid, alkali, or fire
Nitrosyl chloride (ClNO)	Very corrosive and irritating to eyes, skin, and mucous membranes
Nitrosyl fluoride (FNO)	Rocket propellant, highly irritating to eyes, skin, and mucous membranes
Oleum ($H_2SO_4SO_3$)	Also called fuming sulfuric acid, an extremely corrosive substance
Osmium tetroxide (OsO_4)	Irritant to eyes, respiratory tract, and skin
Oxalic acid ($C_2H_2O_4$)	Caustic to mucous membranes, skin, and eyes
Paraquat ($C_{12}H_{14}N_2$)	Nausea, vomiting, diarrhea, and lung tissue injury
Parathion ($C_{10}H_{14}NO_5PS$)	Highly toxic. Causes nausea, vomiting, convulsions, and respiratory failure
Pelargonic acid ($C_9H_{18}O_2$)	Strong irritant
Pentachlorophenol (C_6HCl_5O), also called PCP	Headache, vomiting, convulsions, and heart failure
Perchloryl fluoride ($ClFO_3$)	Oxidizing agent, explosive, can be absorbed through skin
Performic acid (CH_2O)	Irritant to eyes and mucous membranes
Phenarsazine chloride ($C_{12}H_9AsClN$)	Irritant to skin and respiratory tract, tear-gas-like action
Phenol (C_6H_6O)	Ingestion is very toxic, can be fatal due to respiratory failure or cardiac arrest
Phenolsulfonic acid ($C_6H_6O_4S$)	Irritating to skin
Phenylenediamine ($C_6H_8N_2$)	Contact dermatitis, anemia, gastritis, acute poisoning. Leads to respiratory depression in laboratory animals

Phenylhydrazine ($C_6H_8N_2$)   Highly toxic and carcinogenic

Phosgene ($COCl_2$)   Not irritating immediately but quickly fatal, used as a warfare gas. Results from carbon monoxide and nitrosyl chloride

Phosphine (H_3P)   Causes pain in diaphragm, weakness, convulsions, coma, and death

Phosphoric acid (H_3PO_4)   Irritating to eyes, skin, and mucous membranes

Phosphorus oxychloride ($POCl_3$)   Irritating to eyes, skin, and mucous membranes

Phosphorus pentachloride (PCl_5)   Corrosive

Phosphorus pentoxide (P_4O_{10})   Irritant to eyes, skin, and mucous membranes

Phosphorus trichloride (PCl_3)   Irritant to eyes, skin, and mucous membranes

Picloram ($C_6H_3Cl_3N_2O_2$)   Forms tumors in liver and thyroid of laboratory animals

Picoline (C_6H_7N)   Irritant to respiratory tract

Picric acid ($C_6H_3N_3O_7$)   Ingestion results in vomiting, diarrhea, skin eruptions, convulsions, and death

Pinene ($C_{10}H_{16}$)   Absorbed through skin, lungs, and intestines. Inhalation causes palpitations and dizziness. Ingestion can be fatal.

Piperazine ($C_4H_{10}N$)   Corrosive. Ingestion causes nausea, vomiting, and muscle contractions.

Piperidine ($C_5H_{11}N$)   Highly toxic

Potassium arsenate ($KAsO_2HAsO_2$)   Very poisonous in solution. Will reduce silver metal in production of mirrors.

Potassium bifluoride (F_2HK)   Irritating to skin and mucous membranes

Potassium carbonate (KCO_3)   Irritant to skin and eye tissues

Potassium cyanide (KCN)   Poisonous by inhalation. Ingestion can be fatal.

Potassium dichromate ($K_2Cr_2O_7$)   Corrosive to all tissues

Potassium hexafluorosilicate ($K_2(SiF_6)$) Irritant. Ingestion causes vomiting and diarrhea.

Potassium hydroxide (KOH)   Extremely corrosive to tissue

Potassium iodide (KI)   Used to treat goiter and iodine deficiency

Potassium nitrate (KNO_3)   Prolonged exposure produces anemia and kidney disease.

Potassium perchlorate ($KClO_4$)   Explosive

Potassium permanganate ($KMnO_4$)   High concentrations are caustic.

Potassium thiocyanate ($KSCN$)   Skin eruptions and skin disorders

Propanol (C_3H_8O)   Irritating to skin, eyes, and mucous membranes

Propyl acetate ($C_5H_{10}O_2$)   Irritating to skin, eyes, and mucous membranes

Propionic acid (CH_3CH_2COOH)   Irritating to skin, eyes, and mucous membranes

Propylamine ($C_3H_7NH_2$)   Strong irritant and moderately toxic substance

Pyrogallol ($C_6H_6O_3$)   Renal and liver damage, convulsions, circulatory collapse, and death

Pyrethrin ($C_{21}H_{28}O_3$)   Insecticide causes irritation to skin and mucous membranes.

Pyridine (C_5H_5N)   CNS depression, skin irritation, kidney and liver damage

Quinidine ($C_{20}H_{24}N_2O_2$)   Cardiac depressant

Quinine ($C_{20}H_{24}N_2O$)   Plant alkaloid used to treat malaria

Reserpine ($C_{33}H_{40}N_2O_9$)	Sedative, hypotensive, and tranquilizing effect. May be carcinogenic
Rotenone ($C_{23}H_{22}O_6$)	Numbness of oral membranes, nausea, vomiting, respiratory paralysis, and death
Salicylaldehyde ($C_6H_4(OH)CHO$)	A strong skin irritant
Salicylic acid ($C_7H_6O_3$)	Large doses can cause abdominal pain.
Sarin ($C_4H_{10}FPO_2$)	Nerve gas, very toxic
Saxitoxin ($C_{10}H_{17}N_7O_4$)	Extremely toxic. May produce respiratory failure, alkaloid of red tide
Silver acetylide (Ag_2C_2)	Explosive
Silver azide (AgN_3)	Explodes with great violence
Silver fulminate ($Ag_2C_2N_2O$)	Explosive
Silver oxide (AgO)	Irritating to skin, eyes, and mucous membranes
Silvex ($C_9H_7Cl_3O_3$)	Irritating to skin, eyes, and mucous membranes
Soda lime ($NaOH$)	Corrosive and irritating to skin and mucous membranes
Sodium (Na)	Corrosive and irritating to skin and mucous membranes
Sodium arsenate ($NaAsO_2$)	Highly toxic
Sodium azide (NaN_3)	Highly toxic
Sodium borate ($Na_2B_4O_7$)	Vomiting diarrhea, shock, and death
Sodium cacodylate ($C_2H_6AsNaO_2$)	Toxic, nephritis
Sodium cyanide ($NaCN$)	Highly toxic, fatal
Sodium dichromate ($Na_2Cr_2O_7$)	Irritant to skin and mucous membranes
Sodium hydroxide ($NaOH$)	Corrosive and irritating to skin and mucous membranes
Sodium oxalate ($Na_2C_2O_4$)	Intestinal distress, CNS and cardiac depression, death
Sodium peroxide (Na_2O_2)	Corrosive
Soman ($C_7H_{16}FO_2P$)	Nerve gas, very toxic
Sorbic acid ($C_6H_{10}O$)	Irritant to mucous membranes in high doses
Stibine (SbH_3)	Toxic in solution or if dust is inhaled
Strychnine ($C_{21}H_{22}N_2O_2$)	Extremely poisonous
Styrene (C_8H_8)	Irritant to eyes and respiratory membranes
Sulfuric acid (H_2SO_4)	Corrosive to all body tissues
Sulfur tetrafluoride (SF_4)	Irritant and corrosive
Sulfur trioxide (SO_3)	Irritant and corrosive
2,4,5-T (Agent Orange) ($C_8H_5Cl_3O_3$)	Skin irritation, liver degeneration, also toxic to central nervous system
TCDD (see Dioxin)	
Tabun ($C_5H_{11}N_2O_2P$)	Toxic, potent cholinesterase inhibitor, fatal
Tannic acid ($C_{76}H_{52}O_{46}$)	Dye component used as an astringent in veterinary medicine
Tartaric acid ($C_4H_6O_6$)	Concentrated solutions are irritating
Tellurium (Te)	Nausea, vomiting, CNS depression
Tetraethyl lead ($C_8H_{20}Pb$)	May be absorbed through skin or inhaled from gasoline engine exhaust, toxic

Tetralin ($C_{10}H_{12}$)	Irritating to eyes, skin, and mucous membranes
Thalidomide ($C_{13}H_{10}N_2O_4$)	Teratogen, prescription drug used to treat numerous conditions
Thallium (Tl)	Nausea, vomiting, diarrhea, coma, convulsions, death. Many thallium compounds are very toxic.
Thionyl chloride ($SOCl_2$)	Irritating to eyes, skin, and mucous membranes
Thionyl fluoride (SOF_2)	Irritating to eyes, skin, and mucous membranes
Thiourea (CH_4N_2S)	Hepatic tumors reported
Thiram ($C_6H_{12}N_2S_4$)	Irritating to eyes, skin, and mucous membranes
Titanium tetrachloride ($TiCl_4$)	Irritating to eyes and respiratory tract
Tolidine ($C_{14}H_{16}N_2$)	Carcinogenic
Toludine ($C_7H_7NH_2$)	Carcinogenic
Toluene (C_7H_8)	Anemia, narcotic in high concentrations
2,4-Toluenediamine ($C_7H_{10}N_2$)	Carcinogenic
Toluene 2, 4-diisocyanate ($C_9H_6N_2O_2$), (also called TDI)	Irritating to skin and respiratory membranes
Toluenesulfonic acid ($C_7H_8O_3S$)	Irritating to skin and mucous membranes
Toxaphene ($C_{10}H_{10}Cl_8$)	CNS stimulation, convulsions, and death
Tributylamine ($CH_{12}H_{27}NH_2$)	CNS stimulation, skin irritation
Tributyl phosphate ($C_{12}H_{27}O_4P$)	Irritating to mucous membranes
Trichloroacetaldehyde (Cl_3C_2HO)	Mutagenic
Trichloroacetic acid (CCl_3COOH)	Very corrosive
Trichloroacetonitrile (CCl_3CN)	Irritant to eyes, skin, and mucous membranes
Trichlorobenzene ($C_6H_3Cl_3$)	Irritant to eyes, skin, and mucous membranes
Trichloroethane ($C_2H_3Cl_3$)	Irritant to eyes, skin, and mucous membranes
Trichloroethylene (C_2HCl_3)	Headache, fatigue, and visual disturbances. Carcinogenic
Triethylamine ($C_6H_{15}N$)	Irritant to eyes, skin, and mucous membranes
Trimethylene chlorohydrin (C_3H_7PC)	Toxic to central nervous system and gastrointestinal tract
Trinitromethane (($CH-CNO_2)_3$)	Irritant to eyes, skin, and mucous membranes
Tungsten (W)	Inhalation of dust causes coughing, destruction of lung tissue, and some reported deaths.
Uranium (U)	Uranium salts are highly toxic, causing kidney damage and lung cancer.
Urethane (C_3HCl_3)	Solvent of pesticides, also carcinogenic
Valeric acid ($C_5H_{10}O_2$)	Corrosive liquid irritant to eyes and skin. Flammable liquid
Vanadium pentoxide (V_2O_5)	Irritates skin and respiratory membranes
Vinyl bromide (C_2H_3Br)	Inhalation produces anesthesia and kidney damage.
Vinyl chloride (C_2H_3Cl)	Very toxic, liver cancer
Vinyl ethel (($C_2H_3)_2O$)	Anesthetic
Vinylidene chloride ($C_2H_2Cl_2$)	Irritant to skin and mucous membranes. Liver disease
Xylene (C_8H_{10})	Toxic in concentrated forms
Xylyl bromide (C_8H_9Br)	Irritates eyes, causing tear formation. War gas agent.
Zinc chloride ($ZnCl_2$)	Fumes irritate skin and mucous membranes.

Appendix B
A Chemical Research Guide

RESOURCES AND RECOMMENDED READING

The purpose of Appendix B is to provide readers with sources that enable them to conduct further research into specific chemicals or to pursue further study of certain topics that are of interest. Thus we are providing a rather extensive list of books, reference materials, government agency addresses and documents, magazines and periodicals, electronic data, and other sources of information to facilitate such activity.

Whenever an attempt is made to research a chemical, investigators will find that data change regularly and information obtained earlier may no longer be correct. In preparing this appendix, we have attempted to provide information that is as current as possible.

Supposedly, over 70,000 chemicals are in commercial use in our environment, and the truth is that there is very little known about most of them. On the other hand, the number of chemicals that are used in great quantity, that are frequently found at hazardous waste sites, or that appear as products from accidental releases is much smaller.

Thus, there is a great deal of information out there, and readers should be aware that the following list is merely a portion of what is available. Knowing how to research a chemical is a powerful research tool. It is important when you are doing any research to remember the following recommendations:

1. Use the *C A S* number (Chemical Abstract Service Registry Number). This number is assigned to specific elements or chemical compounds by the American Chemical Society's Chemical Abstracts Service. This number conclusively identifies a material regardless of its name. Two organizations that can help you track down a CAS number are (1) The Chemical Manufacturers Association (CHEMTREC) (800) 262-8200 and (2) The American Chemical Society

2. Confirm your information with two or more sources whenever possible.

3. Find local experts who understand chemicals.

4. Always cite your sources with name of book or article, author(s), publisher, city of publisher, and date of publication. Do this for books, articles, videos, and personal communications such as letters, memos, and interviews.

5. Happy Hunting!

BOOKS

Anderson, P. D. 1984. *Basic Human Anatomy and Physiology: Clinical Implications for the Health Professions.* Monterey, CA: Wadsworth Health Sciences Division.

Brauer, R. L. 1990. *Safety and Health for Engineers.* New York: Van Nostrand Reinhold.

Casarett, L. J., Klaassen, C. D., Amdur, M. O., & Doull, J. 1996. *Casarett and Doull's Toxicology: The Basic Science of Poisons,* 5th ed. New York: McGraw-Hill.

Crowley, L. V. 1983. *Introduction to Human Disease.* Monterey, CA: Wadsworth Health Sciences Division.

Hales, D. R. 1989. *An Invitation to Health.* San Francisco: Benjamin/Cummings.

Hunter, D. 1987. *The Diseases of Occupations.* Boston: Little, Brown.

Lawrence, W. W. 1976. *Of Acceptable Risk: Science and the Determination of Safety.* Los Altos, CA: William Kaufmann.

National Safety Council. 1988. *Fundamentals of Industrial Hygiene,* 3rd ed. Chicago: National Safety Council.

National Safety Council. 1955. *Accident Prevention: Environmental Management.* Chicago: National Safety Council.

Ottoboni, M. A. 1991. *The Dose Makes the Poison: A Plain Language Guide to Toxicology,* 2nd ed. New York: Van Nostrand Reinhold.

Williams, P. L., & James, L. B. 1987. *Industrial Toxicology: Safety and Health Applications in the Workplace.* New York: Van Nostrand Reinhold.

Williams, R. J. 1977. *Biochemical Individuality.* Austin, TX: University of Texas.

Sullivan, J. B., & Kreiger, G. R. 1922. *Hazardous Materials Toxicology.* Baltimore: Williams & Wilkins.

Winter, C. K., Seiber, J. N., & Nuckton, C. F. (Eds.). 1990. *Chemicals in the Human Food Chain.* New York: Van Nostrand Reinhold.

REFERENCE MATERIALS

Clayton, G., & Clayton, F. 1981. *Patty's Industrial Hygiene and Toxicology,* 3rd rev. ed. New York: John Wiley.

Coleman, E. J., & Morse, R. A. 1992. *Data: Where It is and How to Get It, The 1993 Directory of Business Environment and Energy Data Sources.* Arnold, MD: Coleman/Morse Associates Ltd.

Driesbach, R. H., & Robertson, W. O. 1987. *Handbook of Poisoning,* 12th ed. Los Altos, CA: Appleton and Lange.

Government Institutes. 1992. *Environmental Telephone Directory.* Rockville, MD: Author.

Hensyl, W. R. 1990. *Stedman's Medical Dictionary.* Baltimore, MD: Williams & Wilkins.

Hodgson, E., Mailman, R. B., & Chambers, J. E. 1994. *Dictionary of Toxicology.* New York: Van Nostrand Reinhold.

Lewis, R. J. 1991. *Hawley's Condensed Chemical Dictionary.* New York: Van Nostrand Reinhold.

Proctor, N. N., & Hughes, J. P. 1984. *Chemical Hazards of the Workplace.* Philadelphia: J. B. Lippincott.

Sax, I. N. 1992. *Dangerous Properties of Industrial Materials,* 8th ed. New York: Van Nostrand Reinhold.

Turkington, C. 1994. *Poisons and Antidotes.* New York: Facts On File Books.

GOVERNMENT PUBLICATIONS

Agency for Toxic Substances and Disease Registry (ATSDR). *Toxicological Profiles.* ATSDR Atlanta, GA: U.S. Public Health Service. Dept of Health and Human Services. PH: (404) 639-0727.

Bureau of the Census. 1994. *Current Industrial Reports: Annual Report on Inorganic Chemicals.* Publication no. MA28A. Washington, DC: Dept. of Commerce.

Bureau of Mines. *Minerals and Mining Subject Bibliography.* Office of Public Mines. Washington, DC: Government Printing Office. PH: (202) 501-9650.

Environmental Protection Agency (EPA). *ACCESS EPA.* Publication No.: EPA 220-B-93-008. Acquire from: GPO, PH: (202) 783-3238, or NTIS, PH: (703) 487-4650.

Environmental Protection Agency (EPA). *Toxics Release Inventory (TRI)*. Washington, DC: OPPT Information Management Division. PH: (202) 260-4655.

Environmental Protection Agency (EPA). 1993. *A Guide to Selected National Environmental Statistics in the U.S. Government*. Center for Environmental Statistics. Washington, DC: OPPT Information Management Division. PH: (202) 260-4655.

Federal Emergency Management Agency (FEMA). *Home Study Courses*. Five courses in emergency planning and preparedness for natural disasters and those caused by human beings. FEMA Homestudy Program, Emergency Mgt. Institute, 16825 So. Seton Ave, Emmittsburg, MD 21727. PH: (301) 447-1162. Also see Learning Resource Center, which offers a lending library. PH: (800) 638-1821.

General Accounting Office (GAO). *Monthly and Yearly Catalog of Reports*. PH: (202) 512-6000.

Government Printing Office (GPO). 710 N. Capitol Street NW, Washington, DC 20401. PH: (202) 512-1800, or Bulletin Board, PH: (202) 512-1387.

Institute for Occupational Safety and Health (NIOSH). 1994. *NIOSH Bookshelf and/or Publications Catalog*. NIOSH Centers for Disease Control. Atlanta, GA 30333. PH: (800) 356-4674.

International Trade Commission. 1992. *Synthetic Organic Chemicals: United States Production and Sales*. Washington, DC: USITC Publication 2720.

National Center for Environmental Publications and Information, 11029 Kenwood Road Bldg #5, Cincinnati, OH 45242.

National Institute for Environmental Health Safety (NIEHS). Internet Home Page is under development in conjunction with the National Toxicology Program. The address will be HTTP:\WWW.NIEHS.NIH.GOV\HOME.HTML, National Clearinghouse, George Meany Center, 1000 Newhampshire, Silver Spring, MD 20903. PH: (919) 541-0217.

National Library of Medicine (NLM). *Online Services Reference Manual for MEDLARS*. Bethesda, MD. PH: (800) 638-8480.

National Network of Libraries of Medicine. PH: (800) 338-7657.

National Research Council. 1991. *Environmental Epidemiology, Volume 1: Public Health and Hazardous Wastes*. Washington, DC: National Academy Press.

National Technical Information Service (NTIS). *Catalog of Products Services*. Springfield, VA 22161. PH: (703) 487-4650.

Occupational Safety Health Administration (OSHA). *OSHA Publications*. Washington, DC 20013-7535. PH: (202) 219-4667.

Office of Technology Assessment (OTA). *Catalog of OTA Publications*. Washington, DC 20510-8025. PH: (202) 224-8996.

NEWSLETTERS, MAGAZINES, AND PERIODICALS

Energy Department of the United States. *ESH SYNERGY*. Office of Environment, Safety and Health. PH: (301) 903-3294.

RACHEL's Environment Health Weekly, Environmental Research Foundation, PO Box 5036, Annapolis, MD 21403. PH: (410) 363-1548.

ELECTRONIC DATA

Access EPA provides information on how to "access" major EPA databases. Washington, DC: OPPT Information Management Division. PH: (202) 260-4655. The following are five such databases:

1. *Online Library System* (OLS) includes several related databases that can be used to locate books, reports, articles, and information on a variety of topics. A user guide is available at (202) 260-2080.

2. *CLU-IN* is a Cleanup Information Bulletin Board that allows you to download newsletter items from "Bioremediation in the Field." It includes treatment technologies. A "User Manual" and "Guided Tour" are available from (513) 891-6685.

3. Vendor Information System for Innovative Treatment Technologies *(VISITT)* has a "User Manual" and a Hotline. PH: (800) 245-4505.

4. *Air CHIEF* provides permitting information in "boiler plate" format and is available through Superintendent of Documents, PO Box 371954, Pittsburgh, PA 1520-7954. PH: (919) 541-5285.

5. Environmental Monitoring Methods Index *EMMI* is an easy-to-use PC database on 2,600 of EPA's regulated chemical substances, methods for their analysis, and regulatory and office-based lists on which they appear. Order from NTIS. PH: (800) 553-6847.

Bureau of the Census publications are available on diskette and on CD-ROM. (See: Bureau of Census, above)

Bureau of National Affairs (BNA) produces an *Environmental Library on CD* that provides a powerful tool for research of environmental data: 1231 25th St NW, Washington, DC 20037. PH: (800) 373-1033.

Department of Transportation sponsors a bulletin board entitled: *Hazardous Materials Information Exchange* with information on training, laws, news, literature, and databases. PH: (312) 972-3275. Setting N-8-1. For assistance, PH: (800) 367-9592. For "Manual," PH: (800) 752-6367.

FEMA Emergency Management Institute operates an interactive electronic bulletin board. PH: (301) 447-6434. Setting N-8-1 or E-7-1 (see: FEMA, above).

Hazardous Substance Data Bank (HSDB). National Library of Medicine (NLM). MEDLARS has over 4,300 chemical substance records. See NLM entry above.

Hazardous Substance Fact Sheets. Numerous databases are available from the Canadian Centre for Occupational Health and Safety (CCOHS) on CD-ROM. The Chem Source c/o CCOHS, 250 Main St East, Hamilton, Ontario, CANADA L8N 1H6. PH: (800) 668-4284.

Hazardous Substance Fact Sheets. New Jersey Department of Health. These fact sheets are available on RACHEL (Remote Access Chemical Hazards Electronic Library). Environmental Research Foundation (ERF), PO Box 5036,

Annapolis, MD 21403-7036. PH: (410) 263-1548, e-mail: erf@rachel.clark.net

National Safety Council distributes a software program entitled *Computer-Aided Management of Emergency Operations* or *CAMEO,* which provides mapping and plume modeling capability coupled with a chemical database. Customer Service, PO Box 558, Itasca, IL. PH: 60143-0558. PH: (800) 621-7619.

National Technical Information Services (NTIS) is the largest single source for public access to federal information and has the information available on compact diskettes, CD-ROMs, and on-line services. Many of the services and databases can be purchased through commercial services. Order at PH: (800) 553-6847.

Occupational Health Services, Inc., distributes software entitled *"OHS MSDS on Disk,"* which contains 96,000 material safety data sheets. 11 W. 42nd St., 12th Floor, New York, NY 10036. PH: (212) 789-3535 or (800) 445-6737.

OSHA maintains a news bulletin board at (202) 219-7316, an electronic bulletin board at (202) 219-4748. Setting 0-8-1. For "Manual," PH: (202) 219-7316. Through the GPO, OSHA has a yearly subscription to the OSHA CD-ROM: OSHA Regulations, Documents, and Technical Information. Order no. 729-013-00000-5. Write Government Printing Office (GPO), 710 N. Capitol Street NW, Washington, DC 20401. PH: (202) 512-1800, or Bulletin Board, PH: (202) 512-1387. Document Files include "Interpretive Quips," a valuable aid to EHS.

Toxics Release Inventory. CD-ROM and high-density diskettes are available from the Government Printing Office, 710 N. Capitol Street NW, Washington, DC 20401. PH: (202) 512-1800, or Bulletin Board, PH: (202) 512-1387 and from the National Technical Information Services (PH: (800) 553-6847. These data are also available through the Right To Know Network (TRK NET). PH: (202) 234-8570. Setting 8-N-1. Log in as "public" or call OMB Watch. PH: (202) 234-8494.

A FEW PRIVATE SOURCES

Austin Associates distributes a software package called *HAZMAX,* which combines MSDS summaries, wastestream reports, materials inventories, VOC reports, and other data for small- to medium-sized businesses: 2880 Eager Rd, LaFayette, NY 13084. PH: (315) 677-7794.

Bowman Environmental Engineering carries a full line of software for air-dispersion modeling called BEE-Line Software: Dallas, TX. PH: (214) 233-5463.

ERM Computer Services distributes software entitled *Enflex Info*, which fully integrates standard management compliance and summary reports and contains the full text of environmental regulations. 855 Springdale Dr., Exton, PA 19341. PH: (800) 544-3118.

GDS Communications distributes a database of over 4,200 chemicals keyed to the Department of Transportation response list and provides essential information for the medical responder and emergency room care for victims of acute chemical contamination: 2380 Riverdale Dr No., Miramar, FL 33025. PH: (800) 982-2566 or (305) 431-2566. E-mail: Douglas.Stutz@internetMCI.COM

J.J. Keller Associates, Inc., distributes both an OSHA Compliance Software package called *Keller-Soft* and a chemical database called Chemical Re-A-Dex. 3033 W. Breezewood Lane, Neenah, WI 54957-0368. PH: (800) 558-5011, Ext 2654.

Jordan Systems, Inc., distributes a software package that cross-references MSDS, RTKs, inventories, and 2,600 chemicals called *Hazardous Materials Manager.* 1901 Broadway, Ste 301, Iowa City, IA 52240. PH: (319) 338-8950.

Logical Technology, Inc., distributes a software system entitled Hazardous Materials Information Network *HAZMIN*, which combines MSDS database management, regulatory listing, and produce process review. 5113 N. Executive Dr., Peoria, IL 61614. PH: (800) 373-6742.

Micromedex, Inc., has developed over 14 databases that include RTECS from NIOSH, HSDB from NLM, IRIS from EPA, OHM-TADS from EPA, CHRIS from the US Coast Guard, DOT Emergency Response Guides, NJ Facts Sheets, and others that are available on one system: *TOMES Plus.* 6200 S. Syracuse Way, Ste 300, Englewood, CA 80111-4740. PH: (800) 525-9083 or (303) 486-6400).

OSHA-Soft, Inc., distributes software modules of pertinent regulations for general industry, construction, SARA, TSCA, and RCRA standards: Amherst, NH. PH: (800) 466-3427.

Regulation Scanning created the first "electronic code book" and has developed *RegScan for Windows*, which provides full-text reproduction of all regulations as published in the CFRs. 30 W. 3rd St., Williamsport, PA 17701. PH: (800) 326-9303.

Resource Consultants, Inc., distributes a database entitled *CHEMTOX*, which matches symptoms with chemical names and first-aid treatment and translates threshold limit values and permissible exposure limits for field personnel. 7121 CrossRoads Blvd., Brentwood, TN 37024. PH: (615) 373-5040.

Solutions Software Corporation distributes the complete U.S. Code of Federal Regulations (50 titles, all full text) and 160,000 MSDS files. Each is available on CD-ROM from 1795 Turtle Hill Rd., Enterprise, FL 32725. PH: (407) 321-7912. E-mail: solution@iag.net

Glossary

Abortifacient An agent that produces an abortion.

Abortus The aborted products of conception.

Abscissa The horizontal axis of a graph.

Absorbent material Material used to soak up liquid (hazardous) material.

Absorption The movement of a chemical from the site of initial contact with the biologic system across a biologic barrier and into either the bloodstream or the lymphatic system.

Acaricide A pesticide used to control spiders, ticks, and mites.

Acclimation The adaptation over several time periods to a marked change in the environment.

Accumulative effect of a chemical The effect of a chemical on a biologic system when the absorption exceeds eliminations, thus increasing the total body burden.

Acetic acid A major component of vinegar; glacial acetic acid is the pure compound.

Acetylcholine An acetic acid ester of choline that serves as a neurotransmitter and neuromuscular transmitter at the synaptic junctions.

Acetylcholinesterase An enzyme present in nerve and muscle that hydrolyzes acetylcholine to choline and acetic acid.

ACGIH American Conference of Governmental Industrial Hygienists, a professional organization that recommends exposure limits (TLVs and BEIs) for toxic substances.

Acid A substance that dissolves in water and releases hydrogen ions (H+). An acid is a proton donor or an electron acceptor. Acids cause irritation, burns, or more serious damage to tissue, depending on the strength of the acid, which is measured by pH: 1 (strongest) to 6 (weakest).

Acid cleaning The use of any acid for the purpose of cleaning materials; methods of acid cleaning are pickling and oxidizing.

Acidity The quantitative capacity of aqueous solutions to react with hydroxyl ions.

Acidosis A condition of decreased pH in the body that may be due to decreased respiratory rate or increased acid consumption.

Active ingredient The chemical that has (pesticidal) action. Active ingredients are listed in order on a label as percentage by weight or as pounds per gallon of concentrate.

Acute effect An adverse effect on a human or animal body that takes place soon after exposure.

Acute poisoning Poisoning by a single exposure to a toxic chemical.

Acute toxicity Any poisonous effect produced by a short-term exposure. The LD_{50} of a substance is typically used as a measure of its acute toxicity.

Additive effect A biological response to an exposure to multiple chemicals that is equal to the sum of the effects of the individual agents.

Administrative control A method of controlling employee exposures to contaminants by job rotations and variations in work assignments.

Adrenal glands A pair of endocrine glands superior to the kidneys. Adrenal glands produce steroids, hormones related to water balance, and norepinephrine.

Adsorption The attachment of the molecules of a liquid or gaseous substance to the surface of a solid.

Adulterant A chemical impurity or substance that, by law, does not belong in a food, plant, animal, or pesticide formulation.

Aerobic A life process that depends on the presence of molecular oxygen.

Aerosol A suspension of a liquid or solid particles in a gas.

Airborne Refers to that which can be carried by or in the air, for example, dusts and mists.

Air pollution The presence of contaminant substances in the air that interfere with human health or air quality; a level of pollutants that exceeds air quality standards, as prescribed by law, and may not be exceeded over a specified period of time or in a defined area.

Air-reactive Possessing the ability to ignite at normal temperatures when exposed to air.

Aldehyde Any of a group of various reactive compounds, such as acetaldehyde, and characterized by the presence of a CHO group.

Alkali Same as a base. *See* Base.

Allergen A substance that causes an allergy.

Allergy Same as hypersensitivity. An allergy is a reaction to a substance that occurs through a change in the immune system caused by the production of antibodies and is usually experienced by only a small number of people exposed to a substance. Allergic reactions in the workplace tend to affect the skin (see dermatitis) and lung (see asthma).

Alpha particle The largest of the common radioactive particles; it is identical to a helium nucleus, travels only 3 to 4 inches, and is stopped by a sheet of paper.

Anesthetic A chemical that causes a total or partial loss of sensation. Overexposure can cause impaired judgment, dizziness, drowsiness, headache, unconsciousness, and even death.

Anhydrous Containing no water.

Animal One of the major kingdoms of living organisms. Multicellular eukaryotes capable of locomotion.

Anion A negatively charged ion in a solution, such as a hydroxyl ion (OH^-).

Anode The positively charged electrode in an electrochemical cell.

ANS *See* Autonomic nervous system.

ANSI American National Standards Institute, a private organization that recommends safe work practices and engineering designs.

Antagonism Situation in which two chemicals interfere with each other's actions such that the net impact is less than the action of either chemical or substance individually.

Aquifer An underground river or lake contained in a bed, or layer, of earth, gravel, or porous storage that contains water.

Aromatic hydrocarbons A class of chemicals consisting of a resonant electron structure, that is, the benzene ring.

Asbestos A widely used mineral, often of a fibrous nature, that can contaminate either air or water and promote cancer after inhalation or ingestion.

Asphyxiant A vapor or gas that can cause loss of consciousness and death due to lack, or reduction, of breathable oxygen.

Asthma Constriction of the airways (bronchial tubes) to the lungs, producing symptoms of cough and shortness of breath. It may be an allergic or an emotional response.

Atmosphere The layer of gas over the surface of a planet. On earth, a mixture of gases, typically 20% oxygen, 80% nitrogen, and assorted trace gases; suitable for breathing and supporting life.

Atmospheric pressure Weight of the air above the earth's surface, typically 14.7 pounds of pressure per square inch.

Atom The smallest particle of an element that can exist while maintaining all of the properties of that element.

Atomic number Number of protons in the nucleus of an atom.

Auto-ignition temperature The temperature to which a closed or nearly closed container must be heated so that the flammable liquid, when introduced into the container, will ignite spontaneously or burn.

Autonomic nervous system (ANS) That portion of the nervous system responsible for "automatic" activities, such as heartbeat, respiration, digestion, blood pressure.

Background level The concentration of a chemical in a natural, unpolluted area; the concentration produced by nature.

Bacteria Typically prokaryotic microorganisms, forming one of the major kingdoms of living organisms.

Baghouse An air pollution abatement device used to trap particulates by filtering gas streams through large fabric bags; similar to a vacuum cleaner bag.

Base A substance that dissolves in water and releases a hydroxyl ion (OH⁻); an electron donor or proton acceptor. It has the ability to neutralize an acid and form a salt. Strong alkalis, pH 14, are irritating and may damage tissue. (*Also see* Caustic).

BEI Biological Exposure Index, the maximum recommended value of a substance in blood, urine, or exhaled air, recommended by the ACGIH.

Beryllium A metal that is hazardous to human health upon absorption. Dust is increased by machining, and it is frequently discharged from foundries and ceramic and propellant plants.

Beta particle An electron, produced by electron tubes (such as television picture tubes) and by radioactive decay. It may cause skin burns; it is stopped by a thin sheet of metal.

Bioaccumulation The retention or accumulation of chemicals in living things, often in a particular part of a living thing, such as a liver or kidney.

Biochemistry All aspects of chemistry that involve living organisms.

Bioconcentration (biomagnification) The tendency of certain elements or chemicals to become concentrated as they move into and up the food chain.

Biodegradable Capable of being decomposed in a relatively short period through the actions of microorganisms.

Biosynthetic A synthetic chemical action requiring energy from the body for reaction.

Biotransformation The transformation process that occurs within a living organism (hence *bio*) of altering exogenous and endogenous chemicals.

BLEVE Boiling Liquid Expanding Vapor Explosion. The explosion that occurs when a container failure releases gas to the atmosphere and releases energy rapidly and violently. For example, a container of liquefied petroleum gas boils when heated and may cause its container to explode.

Blood Circulatory fluid of humans; approximately one-half water with dissolved substances and plasma and one-half a variety of cells and corpuscles (formed elements).

Blow down valve A manually operated valve whose function is to quickly reduce tank pressure to that of atmospheric pressure.

Boiling point The temperature at which a liquid boils and changes rapidly to a vapor (gas) state at a given pressure. (*See* Evaporation.) Expressed in degrees Centigrade (°C) or degrees Fahrenheit (°F) at sea level pressure (760 mm Hg).

British thermal unit (Btu) A unit of measurement of heat; the quantity of heat required to raise the temperature of 0.45 kg (1 pound) of water 1°F at a specified temperature.

Broadcast application In pesticide use, to spread a chemical or substance over an entire area.

Cal/OSHA California Occupational Safety and Health Administration, a state agency in the Department of Industrial Relations that establishes and enforces worker health and safety regulations. It consists of the Division of Occupational Safety and Health (DOSH), the Consultation Service, the Standards Board, and the Appeals Board.

Cancer A condition in which a cell or group of cells proliferates in an unchecked, uncontrolled, or unregulated manner.

Carbon Element with an atomic number of 6 and a nominal atomic weight of 12. Forms the basis of organic compounds and life as we know it.

Carbon dioxide A colorless, odorless, nonpoisonous asphyxiant; normally part of ambient air due to respiration or fossil fuel combustion.

Carbon monoxide A colorless, odorless, poisonous gas produced by incomplete combustion; turns the blood cherry red.

Carbonate A compound containing the carbonic acid (organic acid) group.

Carcinogen A chemical or physical agent that causes or promotes the growth of cancer. Such an agent is often described as carcinogenic. The ability to cause cancer is termed carcinogenicity. Words with similar meaning include *oncogenic* and *tumorigenic.*

Carcinogenesis The process leading to the development of a tumor.

Carcinoma A malignant tumor of epithelial tissue.

Cargo manifest A shipping paper that lists all of the contents being transported by a vehicle or vessel.

CAS number The Chemical Abstracts Service Registry Number is a unique numeric designation that is given to a specific chemical compound to distinguish it from all others; this number may appear on the Material Safety Data Sheet. (*See* MSDS.)

Catabolic reaction The process in which living cells break down substances into simpler substances.

Cathode The negatively charged electrode in an electrochemical cell.

Cation A positively charged ion in a solution; it migrates to the cathode.

Caustic Something alkaline that strongly irritates, corrodes, or destroys living tissue. (*See* Base.)

Caustic soda Sodium hydroxide (NaOH), a strongly alkaline substance used extensively in industrial chemical processing.

Ceiling limit The maximum concentration of a material in air that must never be exceeded, even for an instant.

Cell The structured unit of life, of which tissues are made. There are many types of cells, for example, epithelial cells, connective (blood) cells, nerve cells, and muscle cells. In higher animals each type of cell is specialized to perform a particular function.

CERCLA Comprehensive Environmental Response, Compensation, and Liability Act of 1980. The act requires that the Coast Guard National Response Center be notified in the event of a hazardous substance release. The act also provides for a fund (the Superfund) to be used for the cleanup of abandoned hazardous waste disposal sites.

CFR Code of Federal Regulations. A collection of the regulations that have been promulgated under U.S. law.

Chemical A substance characterized by a definite molecular composition.

Chemical formula A diagram that represents the composition of a substance using symbols to represent each element and subscript numbers showing the number of atoms of each element involved.

Chemical name The scientific designation of a chemical in accordance with the nomenclature system; developed by the International Union of Pure and Applied Chemistry (IUPAC) or the Chemical Abstracts Service's (CAS) rules of nomenclature.

Chemical reduction A chemical reaction in which one or more electrons are transferred to the chemical being reduced from the chemical initiating the transfer (reducing agent).

CHEMTREC Chemical Transportation Emergency Center. Provides assistance during a hazardous materials emergency. It is also a source of information regarding specific chemicals and can contact the manufacturer or other experts for additional information for on-site assistance.

Chlorinated hydrocarbon Organic compound made up of atoms of carbon, chlorine, and usually hydrogen. Chlorinated hydrocarbons include DDT and PCBs; they tend to be very long-lived in the environment, to be toxic, and to accumulate in the food web.

Chlorine A very reactive chemical that readily attaches itself to other elements to form new compounds.

Chlorofluorocarbons (CFCs) Organic compounds made up of atoms of carbon, chlorine, and fluorine, for example, freon.

Chromosome The part of a cell that contains genetic material. (*See* Gene.)

Chronic effect An adverse effect on a human or animal body, such as cancer, which can take months or years to develop after exposure.

Chronic exposure Long-term contact with a substance.

CNS The central nervous system; the brain and spinal cord. Excludes the peripheral nervous system (PNS), which includes the nerves exiting the brain stem and traveling to various locations of the body.

CNS depressant A toxin or substance, such as ethanol or grain alcohol, that affects the CNS by reducing capacity, diminishing sensations, or impairing function.

Combustible Able to catch on fire and burn. The National Fire Protection Association and the U.S. Department of Transportation generally define a "combustible liquid" as having a flash point of 37.8°C (100°F) or higher. (*Also see* Flash point.) *Class A:* ordinary combustibles that leave a residue after burning. *Class B:* flammable liquids and gases. *Class C:* Class A or B fires that occur in or near electrical equipment. *Class D:* combustible metals that are easily oxidized.

Compound A pure substance composed of two or more elements that are chemically combined.

Compressed gas Any gas or mixture of gases having a container pressure exceeding 40 psi at 21.11°C (70°F) or having an absolute pressure exceeding 104 psi at 54.44°C (130°F).

Concentration The relative amount of a specific substance when mixed into a given volume of air or liquid, usually given as parts per million (ppm) or parts per billion (ppb).

Conduction Heat transfer through the movement of atoms within a substance.

Connective One of the four basic tissues; characterized by few cells and large amounts of interstitial material.

Contact dermatitis A skin rash caused by direct contact with an irritating substance.

Contaminants Materials, chemicals, substances, or life forms located where they are not wanted.

Convection Heat transfer from one place to another by actual motion of the heated material.

Corrosive A liquid or solid that causes visible destruction or irreversible alterations in human skin tissue at the site of contact.

Covalent bond A bond involving the sharing of electrons between atoms.

Cubic meter A metric unit of volume, commonly used in expressing concentrations of a chemical in a volume of air. One cubic meter equals 35.3 cubic feet or 1.3 cubic yards. One cubic meter also equals 1,000 liters or one million cubic centimeters.

Curie A measure of radioactivity, 37 billion disintegrations per second.

DDT The first chlorinated hydrocarbon insecticide (1,1,1-trichlorous 2,2-bis (p-chloriphenyl) ethane) It has a shelf-life of 15 years and can collect in fatty tissues of certain animals. In 1972 the EPA banned registration and interstate sale of DDT for all but emergency uses in the United States because of its persistence in the environment and accumulation in the food chain.

Decibel (dB) A unit of sound measurement.

Decomposition The breakdown of a material or substance into simpler parts, compounds, or elements.

Dermal Referring to the skin.

Dermatitis Inflammation of the skin, marked by redness (rash) and often swelling, pain, itching, and cracking. Dermatitis may be caused by an irritant or allergen.

Dioxin Common name for a family of chlorine-containing chemicals, some of which are supremely toxic. Dioxins are the by-products of combustion-based technologies involving chlorine and are known to cause certain types of cancers in humans.

Disease A non-healthy physiological state.

Dose The amount of a chemical that enters or is absorbed by the body. Dose is usually expressed in milligrams of chemical per kilogram of body weight (mg/kg).

Dosimeter An instrument that measures exposure to radiation.

Dust Small particulate material or matter; frequently suspended in air.

Ecology The relationships of living things to one another and to their environment, or the study of such relationships.

Edema A swelling of body tissues due to water or fluid accumulation in tissues.

Electrolyte A liquid, most often a solution, that will conduct an electric current, generally in an electrochemical cell.

Emergency A situation requiring immediate action. (*See* IDLH).

Emulsion A mixture in which one liquid is suspended as tiny drops in another liquid, such as oil in water.

Endothermic reaction A reaction in which heat is absorbed.

Ensemble A group or set, frequently used in referring to a particular configuration of equipment designed or designated for dealing with a specific hazard or hazardous situation.

Environmental Protection Agency (EPA) The United States' government agency responsible for protecting the environment.

EPA Registration Number (EPA Reg. No.) The number that appears on pesticide labels to identify the individual pesticide product.

Epidemiology The study of the patterns of disease and their causes in a population of people.

Epigenesis The development of an organism from an undifferentiated cell, consisting in the successive formation and development of organs and parts that do not preexist in the fertilized egg.

Epigenetic Induction of a heritable alteration in one or more critical genes.

Epithelium One of the four basic tissues; characterized by many cells with little interstitial material.

Ester An organic compound corresponding in structure to a salt in inorganic chemistry. Esters are derived from acids by the exchange of the replaceable hydrogen or an organic alkyl group.

Evaporation The process by which a liquid is changed into a vapor and mixed into the surrounding air.

Evaporation rate The rate at which a liquid is changed to a vapor under standard conditions, usually compared to the rate of another substance that evaporates very quickly.

Exhaust That atmosphere removed from a particular site.

Exothermic reaction A reaction that produces heat.

Explosive A material capable of burning or detonating suddenly and violently. *Class A:* a material or device that presents a maximum hazard through detonation. *Class B:* a material or device that presents a flammable hazard and functions by deflagration. *Class C:* a material or device that contains restricted quantities of either Class A or Class B explosives, or both, but presents a minimum hazard.

Explosive limits The range of concentrations (percent by volume in air) of a flammable gas or vapor that can result in an explosion from ignition in a confined space. Usually given as upper and lower explosive limits. (*See* UEL and LEL.)

Exposure The contact of an organism with a hazard.

Fate The transport and transformation of a pollutant or toxin.

Federal depository library A program administered by the Government Printing Office, which makes selected government publications available to two types of depository libraries: (a) 53 Regional Federal Depository Libraries located mainly at state libraries and state university libraries and (b) 1347 U.S. Government Selective Documents Depositories, one in almost every major metropolitan region.

Ferrous Relating to or containing iron.

Fiber A solid particle whose length is at least three times its width.

Filtrate A liquid that has been passed through a filter.

First law of thermodynamics (energy) In any physical or chemical change, no detectable amount of energy is created or destroyed, but energy can be changed from one form to another.

Fissionable isotope Isotope, such as uranium 235, that can split apart when hit by a neutron at the right speed and thus undergo nuclear fission.

Flammable A substance that catches on fire easily and burns rapidly. The National Fire Protection Agency and the U.S. Department of Transportation define a flammable liquid as having a flash point below 37.8°C (100°F). Same as *inflammable.*

Flammable gas Any compressed gas that will burn.

Flammable limits The range of gas or vapor concentrations (percent by volume in air) that will burn or explode if an ignition source is present. (*See* LEL and UEL.)

Flammable liquid Any liquid having a flash point below 37.78°C (100°F) as determined by tests prescribed in the federal regulations.

Flammable solid Any solid material, other than an explosive, that is liable to cause fires through friction, absorption of moisture, spontaneous chemical changes, retained heat from manufacturing or processing, or that can be ignited readily, and when ignited burns vigorously and persistently.

Flash point The lowest temperature at which a liquid gives off enough flammable vapor to ignite and produce a flame when an ignition source is present.

Fly ash Noncombustible particles carried by flue gas; frequently elementally concentrated due to the combustion process.

Food additive A natural or synthetic chemical deliberately added to processed foods.

Food web Food chain. The relationship of predators and their prey in natural ecosystems: primary producers are eaten by small animals which, in turn, are eaten by larger predators. Together, all of these relationships are called the food web.

Fume Very fine solid particles formed from, typically, recondensed vaporized metals.

Fungi Single-celled or multinucleate organisms such as mushrooms, molds, and yeasts.

Fungicide A pesticide that controls or inhibits fungus growth.

Gamma ray A photon of ionizing energy, typically produced by nuclear reactions; also a high energy X ray.

Gas Third state of matter, composed of diffuse molecules, such as in the Earth's atmosphere.

Gaylor's model A linear graphical extrapolation from zero to the upper confidence level of the lower limit of a dose response curve with a vertical drop down to zero risk. This triangular region encompasses all possible risk and, therefore, eliminates discussion on how to model extrapolations to low dose.

Geiger counter An electrical device that detects the presence of ionizing radiation.

Gene The part of a chromosome that carries a particular inherited characteristic.

Glial cells Nervous system cells responsible for the protection, care, and function of neurons.

Gonads Repository for genetic material; the testes in the male and ovary in the female.

Gram (g) A metric unit of mass. One U.S. ounce equals 28.4 grams; 1 pound equals 454 grams. There are 1000 milligrams (mg) in 1 gram. One kilogram equals 1000 grams (2.2 pounds).

Half-life (a) the period of time in which one-half of the nuclei of a specific radioisotope emits its radiation; (b) the period of time in which one-half of a chemical substance is transformed into a different substance or substances; (c) the period of time in which one-half of a chemical compound is excreted by a living organism or its different tissues or organ systems.

Hard water Alkaline water containing dissolved mineral salts that interfere with some industrial processes and prevent soap from lathering.

Hazard That which may be hazardous to life upon exposure; typically toxic, flammable, explosive, and corrosive materials or unsafe living or working conditions.

Hazardous air pollutant A substance covered by air quality criteria that may cause or contribute to illness or death. Such substances include asbestos, beryllium, mercury, and vinyl chloride.

Hazardous class A group of materials designated by the Department of Transportation (DOT) that share a common major hazardous property, such as flammability, corrosivity, radioactivity.

Hazardous material Any chemical that is a physical hazard or a health hazard.

Hazardous waste A waste material that is classified as toxic, corrosive, flammable, explosive, radioactive, or biological/infectious.

Heart Major organ responsible for pumping blood throughout the body.

Heat of fusion The quantity of heat that must be supplied to a material at its melting point to convert it completely to a liquid at the same temperature.

Heat of vaporization The quantity of heat that must be supplied to a liquid at its boiling point to convert it completely to a gas at the same temperature.

Herbicide A type of pesticide designed to kill a plant or inhibit its growth.

HESIS Hazardous Evaluation System and Information Service; provides information to workers, employers, and health professionals about the health effects of toxic substances and how to use them safely.

Homeostasis A state of body equilibrium or stable internal environment of the body.

Hydrocarbon A compound found in fossil fuels that contains carbon and hydrogen, typically, an organic compound.

Hydrogen sulfide (H_2S) A gas that smells like rotten eggs, which is emitted during organic decomposition; a by-product of oil refining and burning; TWA 10 ppm, IDLH 300 ppm.

Hydrolysis The process in which water is used to split a substance into smaller particles.

Hydroxyl group An oxygen atom and a hydrogen atom bonded together.

Hygiene The science of health and its preservation.

IDLH Immediately Dangerous to Life or Health; a term used to describe an environment that is very hazardous due to a high concentration of toxic chemicals or insufficient oxygen or both.

Igneous rock Rock formed when molten rock material (magma) wells up from the Earth's interior, cools, and solidifies into rock masses.

Ignition temperature The lowest temperature at which a substance will catch on fire and continue to burn.

Incineration The combustion of organic matter or organic waste.

Incompatible Describes materials that could cause dangerous reactions from direct contact with one another.

Industrial hygiene That branch of preventive medicine concerned with the protection of health of the industrial population.

Inflammable Same as *flammable*.

Ingestion The taking in and swallowing of a substance through the mouth.

Inhalation The breathing in of a substance.

Initiation Viewed as the first step of carcinogenesis, involving the induction of an irreversibly altered cell; a mutational event.

Inorganic compound One of two major classes of chemical compounds; it contains no carbon.

Insecticide A type of pesticide designed to kill insect life.

Integrated pest management (IPM) Combined use of biological, chemical, or cultivation methods in proper sequence and timing to avoid economically unacceptable loss of a crop or livestock animals; seeks to minimize use of toxic chemicals (pesticides).

Integument A major organ system forming a barrier between the inside and outside of the body; it includes skin, hair, and nails.

Ion A molecule, atom, or radical that has lost or gained one or more electrons and has, therefore, acquired a net electric charge. Positively charged ions are cations; negatively charged ions are anions. In general, an ion has entirely different properties from the element (atom) from which it was formed.

Ionic bond A bond formed by complete transfer of an electron from one atom to another, which results in ions that are oppositely charged and attract one another.

Irritant A substance that, by contact in sufficient concentration for a sufficient period of time, can cause an inflammatory response or reaction of the eye, skin, or respiratory system.

Isomers Two or more compounds that have the same molecular formula but have different orders of attachment of atoms and thus have different physical and chemical properties.

Kidney A major organ responsible for filtering blood, the regulation of electrolytes, and the excretion of waste via urine.

Kilogram (kg) A metric unit of mass equal to 1,000 grams (2.2 pounds).

Lacrimation Secretion and discharge of tears.

Latency The time between exposure and the first appearance of an effect.

LC_{50} (Lethal concentration 50%) The concentration of a chemical in air that will kill 50% of the test animals inhaling it.

LD (Lethal dose) A concentration of a substance being tested that will kill a test animal.

LD_{50} (Lethal dose 50%) The dose of a chemical that will kill 50% of the test animals receiving it. The chemical may be given by mouth (oral), applied to the skin (dermal), or injected (parenteral). A given chemical will generally show different LD_{50} values depending on how it is given to the animals. It is a rough measure of acute toxicity.

Leachate A material in aqueous solution that occurs as water seeps through (for example, solid waste) and that has the potential of polluting water supplies.

Leaching A process in which various chemicals in upper layers of soil or other solid materials are dissolved and carried into the subsoil and groundwater.

Lead A useful, easily workable, toxic heavy metal element widely used in our society. In humans it causes neurological, reproductive, and growth disorders.

LEL Lower explosive limit or lower flammable limit of a vapor or gas. The lowest concentration that will produce a flash of fire when an ignition source is present. At concentrations lower than the LEL, the mixture is too "lean" to burn.

Lethal Having the ability to kill life.

Ligand The molecules attached to the central atom(s) by coordinate covalent bonds.

Liter The volume of one kilogram of water at 4°C (1.06 quarts); equal to 1000 cubic centimeters.

Litmus A visual indicator of the acidity or alkalinity of various substances.

Local effects Those effects that tend to be limited to the point of contact, for example, poison oak and corrosives such as acids and bases.

LOEL (LOAEL) Lowest observable (adverse) effect level; the lowest dose that produces an observable adverse effect.

Lungs Paired organ responsible for the exchange of gases between the blood and the atmosphere.

Lymphoma Malignancy of the lymphoid tissues.

Medium The environmental vehicle by which a substance (that is, a pollutant) is carried to the site (for example, surface water, soil, or groundwater).

Melting point The temperature at which a solid substance changes to the liquid state.

Metamorphic rock Rock produced when a preexisting rock is subjected to high temperatures, high pressures, chemically active fluids, or a combination of these agents.

Meter A measure of length based on the spectrographic color line of the element krypton; 1 meter equals 39.37 inches.

mg/m^3 A measure of concentration, or the weight of a substance (mg) in a cubic meter of air (m^3); often used to express PELs and TLVs.

Micro Prefix meaning one one-millionth.

Microfiche A sheet of film, a few inches square, containing rows of images that can be read using a microfiche reader.

Milligram (mg) A metric unit of mass. One gram equals 1,000 mg. One ounce equals 28.375 mg.

Mist Liquid particles of various sizes suspended in the atmosphere.

Mitochondria A specialized cell structure that manufactures ATP.

Mixture A combination of substances held together by physical rather than chemical means.

mm Hg Millimeters (mm) of the metal mercury (Hg); a unit of measurement for pressure. At sea level, the Earth's atmosphere exerts 760 mm Hg of pressure.

Molecular weight The sum of the atomic weights of the atoms in a molecule.

Molecule Combination of two or more atoms of one or more chemical elements held together by chemical bonds. Smallest particle of a compound capable of having the properties of the compound.

Monomer (*See* polymerization).

mp/kg A way of expressing dose; milligrams of a substance (mg) per kilogram (kg) of body weight. (*See* dose.)

MSDS Material Safety Data Sheet. Contains specific health and safety information required by the Federal Hazard Communication Standard for any hazardous substance. The format will vary depending on the manufacturer or supplier.

MSHA Mine Safety and Health Administration; an agency in the U.S. Department of Labor that regulates safety and health in the mining industry. This agency also tests and certifies respirators. (*See* NIOSH).

Mucous membrane The moist, soft lining of the nose, mouth, throat, bronchus, and eyes.

Muscle One of the four basic tissue types, characterized by a fundamental ability to contract.

Mutagen A chemical or physical agent able to change (mutate) the genetic material in cells.

Mutagenesis A process leading to the alteration of DNA.

Mutation An inheritable change in the kind, structure, sequence, or number of component parts of a cell's DNA.

NAAQS National Ambient Air Quality Standards. Maximum allowable level, averaged over a specific period of time, for a certain pollutant in outdoor air.

Necrosis The death of tissue, usually as individual cells, groups of cells, or in small localized areas.

Nervous One of the four basic tissues, characterized by the ability to conduct impulses.

Neurons Family of cells of nervous tissue responsible for the propagation of neural impulses.

Neutralization The chemical addition of either an acid or base to a solution to adjust the pH to 7.0.

NFPA National Fire Protection Association. NFPA has developed a scale for rating the severity of fire, reactivity, and health hazards of substances. References to these ratings frequently appear on MSDSs.

NIOSH National Institute for Occupational Safety and Health, a federal agency that conducts research on occupational safety and health questions and recommends new standards to federal OSHA. NIOSH, along with MSHA, tests and certifies respirators.

NOEL (NOAEL) No-observable-effects level; maximum dose at which no adverse effect is observable.

Noise pollution Any unwanted, disturbing, or harmful sound that impairs or interferes with hearing, causes stress, hampers concentration and work efficiency, or causes accidents.

Nondegradable pollutant Material (such as lead) that is not broken down by natural processes.

Nonpoint source Source of pollution from human activities that occurs on large or dispersed land areas such as cropfields, streets, and lawns. These sources discharge pollutants into the environment over a large area.

Nonrenewable resource Resource that exists in a fixed amount (stock) in the Earth's crust and has the potential for renewal only by geological, physical, and chemical processes taking place over millions to billions of years.

Non-routine task A predictable task that occurs infrequently.

Nucleus The center of an atom, making up most of the atom's mass. The nucleus contains one or more positively charged protons and one or more neutrons with no electrical charge.

Nutrient Any food or element that an organism requires to live, grow, or reproduce.

Odor threshold The lowest concentration of a substance in air that can be smelled. For a given chemical, different people usually have very different odor thresholds.

Oncogenic Tending to cause tumors, whether benign or malignant.

Organic compound One of two major classes of chemical compounds; contains carbon.

OSHA Federal Occupational Safety and Health Administration, an agency in the U.S. Department of Labor that establishes workplace safety and health regulations. Many states, including California, have their own OSHA programs. State OSHA programs are monitored by federal OSHA to ensure they are "at least as effective as" the federal OSHA program.

Oxidant A substance that supplies oxygen for chemical reactions; oxidation is the opposite of reduction.

Oxidation-reduction The process of substances combining with oxygen.

Oxidizer A chemical other than a blasting agent or explosive that initiates or promotes combustion in other materials, thereby causing fire.

Ozone layer Layer of gaseous ozone (O_3) in the stratosphere, 17.7 to 48.3 km (11 to 30 miles) above the surface of the Earth, that filters out harmful ultraviolet radiation from the sun.

Particulates Fine liquid or solid particles, such as dust, smoke, mist, fumes, or smog, found in the air or point source emissions.

Pathogen An organism (for example, certain bacteria and viruses) that produces disease.

Pathway The flow of a substance through a series of reactions or operations.

PCB Polychlorinated biphenyl; a group of toxic, persistent chemicals used in transformers and capacitors. Its sale or use was banned in the United States in 1979.

PEL Permissible exposure level, a maximum allowable exposure level under OSHA regulations.

Persistent pesticides Pesticides that do not break down chemically and remain in the environment after a growing season.

Pesticides A class of chemicals designed to kill or inhibit life forms that humans consider to be pests. Pesticides include fungicides, herbicides, insecticides, and rodenticides.

Petrochemicals Chemicals obtained by refining (distilling) crude petroleum (oil) and used as raw materials in the manufacture of most industrial chemicals, fertilizers, pesticides, plastics, synthetic fibers, paints, medicines, and other products.

pH Indicates how acidic or basic a solution or chemical is, on a scale of 1 to 14. For example, a pH of 1 indicates a strongly acidic solution, a pH of 7 indicates a neutral solution, and a pH of 14 indicates a strongly alkaline solution.

Physical hazard A chemical for which there is scientifically valid evidence that it is a combustible liquid, a compressed gas, an explosive or flammable substance, an organic peroxide, an oxidizer, or a pyrophoric, unstable (reaction), or water-reactive substance.

PNS Peripheral nervous system; the distal portion of the nervous system from the spinal cord to the muscle, motor, and skin sensors.

Point source A stationary location where discharge occurs, such as a smokestack.

Poison A chemical or substance that, in relatively small amounts, produces injury by chemical or biochemical actions when it comes into contact with a susceptible cell or tissue. *Class A:* a gas or liquid that is so poisonous that, when even small amounts are mixed with air, is dangerous to life and health. *Class B:* any substance known to be so toxic to humans that it poses a severe health hazard during transportation.

Pollution Substance introduced into air, water, soil, or food that is not normally present (or not normally present in such high concentrations) and that can adversely affect the health, survival, or activities of humans or other living organisms.

Polymerization A chemical reaction in which small molecules (monomers) combine to form much larger molecules (polymers). A hazardous polymerization is a reaction that occurs at a fast rate and releases large amounts of energy. Many monomers are hazardous in the liquid and vapor states but form much less hazardous polymers. An example is vinyl chloride monomer, which causes cancer but forms the relatively nontoxic polyvinyl chloride (PVC) plastic.

Polyuria The passage of a large volume of urine in a given period, a characteristic of diabetes.

ppb Parts per billion, a measure of concentration, such as parts of a given chemical per billion parts of air or water. One thousand ppb equals 1 ppm.

ppm Parts per million; a measure of concentration, such as parts of a given substance per million parts of air or water. PELs and TLVs are often expressed in ppm.

Precipitate An insoluble material produced by chemical reaction in a solution.

Priority pollutant The Clean Water Act amendments of 1977 listed 126 "priority pollutants" based on criteria of toxicity, persistence, and potential for exposure of living things.

Progression The process whereby a benign tumor becomes malignant due to additional heritable changes to the initiated cells.

Promotion Viewed as the second step of carcinogenesis. The experimentally defined process by which the initiated cell clonally expands into a visible tumor, often a benign lesion; a reversible event.

Proton Positively charged particle in the nucleus of every atom. Each proton has a relative mass of 1 and a single positive charge.

psi Pounds per square inch. A unit of pressure. At sea level, the Earth's atmosphere exerts 14.7 psi, which equals 760 mm Hg.

Pulmonary edema The filling of the lungs with fluid, which produces coughing and difficulty breathing and may cause drowning or death by suffocation.

Pyrophoric Capable of igniting spontaneously when exposed to dry or moist air at or below 54.44°C (130°F).

Rad A unit of measurement of any kind of radiation absorbed by tissue.

Radiation The emission or transfer of energy by photons or particles; photons or particles released by nuclear reactions.

Reaction A chemical transformation or change.

Reactivity The ability of a substance to undergo a chemical reaction (such as combining with another substance). Substances with high reactivity are often quite hazardous.

Reproduction Production of offspring by parents.

Respirator A device worn to prevent inhalation of hazardous substances.

Risk The probability that something undesirable will happen from deliberate or accidental exposure to a hazard.

Risk analysis Identifying hazards and evaluating the nature and severity of risks (risk assessment) to make decisions about allowing, reducing, increasing, or eliminating risks (risk management) and communicating information about risks to decision makers or the public (risk communication).

Risk assessment The process of gathering data and making assumptions to estimate short- and long-term harmful effects or "no" effects on human health or the environment from exposure to hazards associated with the use of a particular product or technology.

Risk-benefit analysis Estimate of the short- and long-term risks and benefits of using a particular product or technology.

Risk communication Communicating information or perspectives about risks to decision makers or the public.

Risk management Using risk assessment and other information to make decisions about reducing, eliminating, introducing, continuing, or increasing risks.

Rodenticide A type of pesticide designed to kill rodents.

Route of entry The means by which material may gain access to the body, for example, inhalation, ingestion, and skin contact.

Safe exposure level The level of exposure that will not result in a health hazard.

Salt The compound formed when the hydrogen of an acid is replaced by metal, or its equivalent, for example, ammonium, NH_4. Generally, the reaction of an acid and a base yields a salt and water; salts tend to ionize (dissolve) in water solutions.

Sarcoma A malignant tumor of the connective tissue.

SCBA Self-contained breathing apparatus. A respiratory protection device that consists of a supply of oxygen, or oxygen-generating material, carried by the wearer.

Secondary pollutant A harmful chemical formed in the atmosphere when a primary air pollutant reacts with normal air components or with other air pollutants.

Second law of thermodynamics In any conversion of heat energy to useful work, some of the initial energy input is always degraded to lower-quality, more dispersed (higher entropy), less useful energy, usually low-temperature heat that flows into the environment.

Sedimentary rock A rock that forms from deposition of eroded materials and in some cases from the compacted shells, skeletons, or other remains of dead organisms.

Shale oil A slow-flowing, dark brown, heavy oil obtained when kerogen (solid, waxy mixture of hydrocarbons) in oil shale (a rock formation) is vaporized at high temperatures and then condensed.

Skin absorption Ability of some hazardous chemicals to pass directly through the skin and enter the bloodstream.

Smog Air pollution associated with smoke and fog; excess criteria air pollutants in the atmosphere.

Smoke Particles suspended in air after incomplete combustion of materials containing carbon.

Soil Complex mixture of inorganic minerals (clay, silt, pebbles, and sand), decaying organic matter, water, air, and living organisms.

Solubility The degree to which a chemical can dissolve in a solvent, forming a solution.

Solute A dissolved substance.

Solution A mixture in which the components are uniformly dispersed. All solutions consist of some kind of a solvent (such as water or other liquid) that dissolves another substance, usually a solid.

Solvent A substance, usually a liquid, capable of dissolving another.

Special fire-fighting procedures Special procedures and personal protective equipment that are necessary when a particular substance is involved in a fire.

Species A group of organisms that resemble one another in appearance, behavior, chemical makeup and processes, and genetic structure.

Specific gravity The ratio of the weight of a volume of the product to the weight of an equal volume of water.

Stability An expression of the ability of a material to remain unchanged under expected and reasonable conditions of storage and use.

STEL Short-term exposure limit; a term used by ACGIH to indicate the maximum average concentration allowed for a continuous 15-minute exposure period.

Stereoselectivity In simple terms, when a molecule has a pair of isomers (mirror-image molecules that are alike in many ways, such as melting and boiling points and solubility, yet are different) that are present in the mixture in equal amounts (a racemic mixture), then metabolism occurs at different rates for each isomer; while one may appear largely unchanged and still be foreign when found in the urine, the other will appear as a fully transformed metabolite.

Subatomic particles Extremely small particles—electrons, protons, and neutrons—that make up the internal structure of atoms.

Sublimation Passing directly from the solid to the vapor state, for example, dry ice.

Substrate An underlying reactant attached to an intermediate product of biotransformation.

Superfund sites Contaminated sites, the cleanup of which was authorized by CERCLA.

Synergistic effect A biological response to dosage of multiple substances in which the net effects are greater than the sum of the effects of the individual agents.

Synthetic chemicals Chemicals that people make, rather than those that nature makes; they tend to be long-lived in the environment.

Synthetic reaction A chemical action in which larger molecules are formed from simpler ones.

Systemic effects Effects that are due to absorption and distribution of an agent into and throughout the body, that is, away from the site of contact. They impact one or more of the 10 basic systems of which humans are composed.

Technical pesticide A highly concentrated pesticide that is intended to combine with other materials to formulate pesticide products.

Teratogen A chemical or physical agent that can lead to malformations in the fetus and birth defects in children (live-born offspring). The ability to cause birth defects is termed *teratogenicity.*

Threshold The lowest dose of a chemical at which a specific measurable (observable) effect occurs.

Time series A set of statistics that are compiled and reported over a period of time.

TLV Threshold limit value, an exposure limit recommended by the ACGIH.

Toxicity The degree of danger to life posed by a substance.

Toxicology The study of the harmful effects of chemicals on living things.

Toxic substance Any substance that can cause acute or chronic injury to the human body or that is suspected of being able to cause diseases or injury under some conditions.

Toxin (*See* Toxic substance).

Toxon A toxic substance within a living organism that will react with it at some specific target (site, enzyme, or cell). This reaction induces a toxic effect.

Trace element An element essential to plant and animal nutrition in trace amounts, 1,000 ppm or less.

Trade name The trademark name or commercial name given to a material by the manufacturer or supplier.

Trade secret Any confidential formula pattern, process, device, information, or compilation of information that is used in an employer's business, and that gives the employer an opportunity to obtain an advantage over competitors who do not know about or use it.

Transformation The chemical alteration of a compound by breakdown into component elements, conversion into other molecules, or reaction with other compounds.

Transport The movement of a substance along a pathway.

TWA Time-weighted average. The average concentration of a chemical in air over the total exposure time, usually an eight-hour work day.

UEL Upper explosive limit. (*See* Explosive limits.)

Unstable Describes a material or substance that is capable of undergoing rapid chemical change or decomposition.

Vapor The gaseous form of a substance that is primarily a liquid or solid at standard pressure and temperature.

Vapor pressure A measure of the tendency of a liquid to evaporate and become a gas. The pressure exerted by a saturated vapor above its own liquid in a closed container at given conditions of temperature and pressure, usually expressed in mm Hg; the higher the vapor pressure, the greater the tendency of the substance to evaporate. (*See* Evaporation, mm Hg, and Volatility).

Volatility A measure of how quickly a substance forms vapors at ordinary temperatures. The more volatile a substance is, the faster it evaporates, and the higher the concentration of vapor (gas) in the air.

Waste disposal methods Proper disposal methods for contaminated material, recovered liquids or solids and their containers.

Water pollution The introduction into water of a substance that is not normally present (or not normally present in such a high concentration) and that is potentially toxic or otherwise undesirable.

Water reactive A chemical that reacts with water to release a gas that is either flammable or presents a health hazard. Also denotes danger when wet.

Wettable powder A finely ground pesticide dust that will mix with water to form a suspension for application. Although this formulation may not burn, it may release toxic fumes when exposed to fire.

Wood preservative A pesticide used to treat any wood product to prevent damage by pests or dry rot.

Index

ISBN 0-02-389551-9